S0-DJC-605

CHICAGO PUBLIC LIBRARY
HAROLD WASHINGTON LIBRARY CENTER
R0006700056

SWARF AND MACHINE TOOLS

SWARF AND MACHINE TOOLS

A guide to the methods used in the handling and treatment of swarf and cutting fluids

Edited by P. J. C. GOUGH

HUTCHINSON OF LONDON

HUTCHINSON & CO (*Publishers*) LTD
178–202 Great Portland Street, London W1

London Melbourne Sydney
Auckland Johannesburg Cape Town
and agencies throughout the world

First published 1970

This book has been compiled by The Machine Tool Industry Research Association from information supplied by the many firms and organisations who have co-operated in its production. The book has been prepared under the guidance of a Steering Committee, the members of which were:
P. J. C. Gough (Chairman), A. C. Abbott, E. G. Allsop, P. D. Beard, H. H. Davies, S. J. Ellis, G. S. Henderson, P. M. Holmes, T. G. Murray, B. Wade, R. F. Worlidge.

This book has been produced by Hutchinson Benham.
It has been set in Imprint type, printed in Great Britain
on smooth wove paper by Anchor Press, and
bound by Wm. Brendon, both of Tiptree, Essex

ISBN 0 09 102240 1

CONTENTS

ACKNOWLEDGEMENTS

The Machine Tool Industry Research Association is indebted to the following persons, and the companies they represent, for their valuable assistance in providing much of the material contained in this volume. In recording our sincere thanks to them, we freely acknowledge that without their expert guidance based on their extensive experience together with their constructive comments, criticisms and technical contributions, this manual would have been incomplete.

STEERING COMMITTEE

A. C. Abbott, The Ford Motor Co. Ltd., Dagenham, Essex; P. D. Beard, Alfa-Laval Co. Ltd., Great West Road, Brentford, Middlesex; H. H. Davies, The Darrold Engineering Co. Ltd., Balds Lane, Lye, Stourbridge, Worcestershire; S. J. Ellis, New Conveyor Co. Ltd., Argyle Works, Stevenage, Herts; G. S. Henderson, Staveley Machine Tools—Archdale Division, Blackpole Works, Worcester; P. M. Holmes, Shell-Mex & B.P. Ltd., Shell-Mex House, Strand, London, W.C.2; T. G. Murray, Philips Filtration Ltd., Blenheim Gardens, Brixton, London, S.W.2; B. Wade, Spencer & Halstead Ltd., Bridge Works, Ossett, Yorkshire; R. F. Worlidge, Euroflow Systems Ltd., Strand House, Strand Street, Poole, Dorset

WORKING PARTIES

Centrifugal separators and hydrocyclones

P. D. Beard (Chairman), Alfa-Laval Co. Ltd., Great West Road, Brentford, Middlesex; R. Golding, Westphalia Separator (G.B.) Ltd., Old Wolverton Road, Wolverton, Bucks; N. W. J. Henney, Ferguson & Timpson Ltd., 142–4 Minories, London, E.C.3; F. T. Morris, Gaston E. Marbaix Ltd., Bessemer Road, Basingstoke, Hampshire; T. B. Povey, Broadwell Engineering Ltd., Dudley Road East, Tipton, Staffordshire; G. C. Waddington, Machine Shop Equipment Ltd., Manor Royal, Crawley, Sussex; A. J. Weeden, Pensalt Ltd., Sharples Division, Tower Works, Doman Road, Camberley, Surrey

Cutting fluids

P. M. Holmes (Chairman), Shell-Mex & B.P. Ltd., Shell-Mex House, Strand, London, W.C.2; G. A. Drake, Alexander Duckham & Co. Ltd., Summit House, Glebe Way, West Wickham, Kent; F. A. Fiddey, Gulf Oil (G.B.) Ltd., 6 Grosvenor Place, London, S.W.1; N. Gullick, Regent Oil Co. Ltd., 117 Park Street, London, W.1; L. H. Haygreen, Esso Petroleum Ltd., Victoria Street, London, S.W.1; F. Hilton, Burmah-Castrol Industrial Ltd., Hyde, Cheshire; E. S. Jones, Edgar Vaughan & Co. Ltd., Legge Street, Birmingham, 4; J. B. Lewis, The Mobil Oil Co. Ltd., 54–60 Victoria Street, London, S.W.1; R. A. Nicholson, Gulf Oil (G.B.) Ltd., 6 Grosvenor Place, London, S.W.1; B. Sismey, Shell-Mex & B.P. Ltd., Shell-Mex House, Strand, London, W.C.2; J. A. Wells, Germ Lubricants Ltd., Bloom Street, Salford, 3, Lancashire

Filters

T. G. Murray (Chairman), Philips Filtration Ltd., Blenheim Gardens, Brixton, London, S.W.2; P. D. Beard, Alfa-Laval Co.

Ltd., Great West Road, Brentford, Middlesex; R. G. Carroll, Hydromation Engineering Co. (G.B.) Ltd., Blandford Road, Poole, Dorset; M. G. Deering, Centri-Spray Ltd., Green Dragon House, 64–70 High Street, Croydon, Surrey; C. W. Ferguson, Swinney Brothers Ltd., Morpeth, Northumberland; D. V. Grantham, Belisle Filtration Ltd., Bel Works, Ridgway Road, Luton, Bedfordshire; E. J. C. Holland, Machine Shop Equipment Ltd., Manor Royal, Crawley, Sussex; I. Kinloch, Belisle Filtration Ltd., Bel Works, Ridgway Road, Luton, Bedfordshire; F. T. Morris, Gaston E. Marbaix Ltd., Bessemer Road, Basingstoke, Hampshire; R. Pizzey, Alfa-Laval Co. Ltd., Great West Road, Brentford, Middlesex; D. H. Russell, Auto-Klean Strainers Ltd., Lascar Works, Staines Road, Hounslow, Middlesex; J. Steer, Vokes Ltd., Henley Park, Guildford, Surrey; P. R. Sylvester, Philips Filtration Ltd., Blenheim Gardens, Brixton, London, S.W.2; A. T. Tyndall, Alfa-Laval Co. Ltd., Great West Road, Brentford, Middlesex; H. W. Whitter, Vokes Ltd., Henley Park, Guildford, Surrey

Hydraulic conveying systems

R. F. Worlidge, Euroflow Systems Ltd., Strand House, Strand Street, Poole, Dorset; R. G. Carroll, Hydromation Engineering Co. (G.B.) Ltd., Blandford Road, Poole, Dorest

Machine-tool manufacturers

G. S. Henderson (Chairman), Staveley Machine Tools—Archdale Division, Blackpole Works, Worcester; H. F. Alder, Kearney & Trecker Ltd., Portland Road, Hove, 3, Sussex; B. Burton, Heald Machines Ltd., Dells Lane, Biggleswade, Bedfordshire; K. G. Hubbard, Vaughan Renault Machine Tools (U.K.) Ltd., Harlescott Lane, Shrewsbury; T. E. Lindem, Herbert-Ingersoll Ltd., Daventry, Northants; R. H. Lyons, Warner-Swasey Asquith Ltd., Water Lane, Halifax; D. Rowe, Churchill-Redman Ltd., Parkinson Lane, Halifax,

Yorks; E. F. Sutton, Cincinnati Milling Machines Ltd., Kingsbury Road, Tyburn, Birmingham, 24

Machine-tool users

A. C. Abbott (Chairman), The Ford Motor Co. Ltd., Dagenham, Essex; G. W. Cutler, British Leyland Motor Corporation, Longbridge, Birmingham; T. E. A. Dicken, Royal Ordnance Factory, Nottingham; H. Fowler, British Leyland Motor Corporation, Leyland, Lancashire; L. J. Hoefkins, Automotive Products Co. Ltd., Tachbrook Road, Leamington Spa, Warwickshire; G. B. Horsler, Skefko Ball Bearing Co. Ltd., Leagrove Road, Luton, Bedfordshire; C. Howard, A.E.I. (Power Group), Trafford Park, Manchester; J. F. Hughes, The Rover Car Co. Ltd., Meteor Works, Lode Lane, Solihull, Warwickshire; T. A. Jinks, Hawker-Siddeley Aviation Ltd., Kingston-upon-Thames, Surrey; A. R. Kedwards, Joseph Lucas (Industries) Ltd., Great King Street, Birmingham, 19; J. W. Rose, English Electric Ltd., Stafford

Mechanical conveying systems

H. H. Davies (Chairman), The Darrold Engineering Co. Ltd., Balds Lane, Lye, Stourbridge, Worcestershire; R. G. Carroll, Hydromation Engineering Co. (U.K.) Ltd., Blandford Road, Poole, Dorset; C. Derby, Broadwell Engineering Ltd., Dudley Road East, Tipton, Staffordshire; F. F. Dyer, Bagshawe & Co. Ltd., Dunstable, Bedfordshire; W. A. Morrison, Moffat & Bell Ltd., 31–37 The Broadway, London, S.W.19; W. F. Pickstone, New Conveyor Co. Ltd., Argyle Works, Stevenage, Hertfordshire; T. B. Povey, Broadwell Engineering Ltd., Dudley Road East, Tipton, Staffordshire; A. J. Tyrrell, Philips Filtration Ltd. Blenheim Gardens, Brixton, London, S.W.2; H. W. D. Wilson, Integrated Conveyors Ltd., Oak Street, Cradley Heath, Warley, Worcestershire

Pneumatic conveying systems

B. Wade (Chairman), Spencer & Halstead Ltd., Bridge Works, Ossett, Yorkshire; D. Mann, Dustraction Ltd., Mandervell Road, Oadby, Leicester; G. Punch, American Air Filter (G.B.) Ltd., Bassingdon Industrial Estate, Cramlington, Northumberland; H. R. Scrivens, Midland Heating & Ventilation Co. Ltd., Redhill Road, Birmingham, 25; P. Swift, Dust Control Equipment Ltd., Thurmaston, Leicester; G. E. Thomas, Gwyn Thomas Ltd., 15 Highfield Road, Edgbaston, Birmingham, 15; A. F. Williams, B.V.C. Engineering Ltd., Leatherhead, Surrey

Swarf-processing equipment

S. J. Ellis (Chairman), New Conveyor Co. Ltd., Argyle Works, Stevenage, Hertfordshire; W. G. Fraser, Manlove-Alliott & Co. Ltd., Nottingham; W. Freeland, Philips Filtration Ltd., Blenheim Gardens, Brixton, London, S.W.2; D. Hall, British Jeffrey-Diamond Ltd., Wakefield, Yorkshire; E. J. C. Holland, Machine Shop Equipment Ltd., Manor Royal, Crawley, Sussex; M. Kemp, Broadwell Engineering Ltd., Dudley Road East, Tipton, Staffordshire; P. B. Ogilvie, Christy & Norris Ltd., Broomfield Road, Chelmsford, Essex; B. Sismey, Shell-Mex & B.P. Ltd., Shell-Mex House, Strand, London, W.C.2; A. T. Tyndall, Alfa-Laval Co. Ltd., Great West Road, Brentford, Middlesex; B. Wade, Spencer & Halstead Ltd., Bridge Works, Ossett, Yorkshire

PREFACE

The continuous and rapid improvement in productivity made possible by modern machine tools capable of high rates of metal removal has created a need for more thought to be given to the problem of dealing with the greatly increased volume of the swarf that is produced. The Machine Tool Industry Research Association (M.T.I.R.A.) recognised that there was a need for work to be carried out in the field of swarf removal and handling and a research programme was formulated. In the early stages of the investigation a small committee of representatives of the machine-tool industry was formed to advise and to study the progress of the work.

It soon became evident that lack of communication between the various parties involved—machine-tool makers, machine-tool users and makers of swarf-handling equipment—was responsible for many of the difficulties being encountered by users of machine tools, and it was decided that the attention of engineers in general should be drawn to the problems and the methods available for their solution at a one-day conference organised by M.T.I.R.A. and held at the University of Aston, Birmingham, on 24th May 1967. The conference was so well attended and the interest shown was so great that M.T.I.R.A. was encouraged to continue its investigations and to make plans to bring together in more permanent form the knowledge and expertise that was known to exist so that all of the parties concerned could be aware of the others' points of view. It was thought that the best way in which this could be done would be to produce a manual that would contain information about the type of equipment that is available for dealing with swarf, offer advice on the methods to be employed with various forms of swarf and which would act as a general guide to engineers interested in the treatment of swarf and cutting fluids.

For the preparation of the book a series of working parties was formed with members drawn from firms whose products are directly associated with some aspect of the subject, and these working parties operated under the guidance of a Steering Committee. M.T.I.R.A. wishes to thank all those persons and firms who have co-operated in the production of this book which it is hoped will be of value to all who are concerned in any way with the handling of swarf.

For the benefit of future editions, indications of new developments, additional information gained from practical experience and other relevant comments will be welcomed by M.T.I.R.A.

1 SWARF AND MACHINE TOOLS

This manual is concerned with a subject that has been regarded—one might almost say disregarded—for very many years as a simple everyday operation in every workshop in which machine tools are used. However, the days when swarf removal was a matter for a man with a rake, a brush, a shovel and a barrow have now disappeared from all but small or medium-sized establishments. Engendered by the need to produce large quantities of war material between 1939 and 1945, and to cope with subsequent peace-time world demands, it has been necessary for manufacturers continually to develop their machines and to advance tooling techniques in order to produce more and more workpieces in less and less time and, consequently, to make swarf at a greater rate. In consequence, the simple operation of the past has, by virtue of the large volumes of swarf now involved, grown into a major problem confronting all sizable establishments and warranting the deepest consideration.

Not only has the removal of greatly increased quantities of swarf from the machine tools to be considered, but it is also increasingly necessary to study the means by which such swarf can be separated from any cutting fluids that may have been used; the treatment of such fluids before re-circulation; the disposal of contaminated fluids or sludge without contravening any local or national law or bye-law; and the presentation of the swarf to the metal merchant in the form most suitable for him and for which he will pay the highest price.

Neglect or mis-handling of swarf may bring other problems into being and can be responsible for:

injury to operators from the sharp edges and retained heat in the cuttings.

skin and other diseases for the machine operators from polluted cutting fluids.

damage to the table surfaces, slideways, leadscrews, bearings and the measuring equipment of machine tools;

damage to the finished surfaces of machined workpieces;

the trapping of particles in places such as blind holes that could cause tool breakages during subsequent operations;

incorrect positioning of workpieces in jigs or fixtures, leading to sub-quality or rejected components.

In the past, it has always appeared that a wide gulf existed between the machine-tool manufacturer and the ultimate user on questions such as are now being considered, and this has led to a feeling that there is a state of split responsibility, probably caused by a lack of communication between the two parties. However, many if not all of the problems involved in removing swarf from machine tools can be solved if there is adequate discussion beforehand between all the parties concerned—machine-tool user, machine-tool manufacturer and the manufacturers of swarf-handling equipment—and the main objective of this manual is to present the points of view of all the parties involved so that none need be ignorant of the contribution that the other can make and of the steps which each must take to facilitate the work of the others. With such considerations in mind, it will readily be seen that the subject is complex and presents a problem that requires careful analysis if an effective and economic solution is to be found.

In making a decision on the type of swarf-handling equipment required many different factors must be taken into account. General solutions to the problem are seldom applicable, and in most effective applications of swarf-handling equipment a unique combination is required.

Before attempting to deal with the removal and treatment of swarf from a workshop, it is necessary to consider the factors that can have an important bearing on the final design and the economics of any system that is adopted and, in fact, on the economics of manufacture as a whole. Some of the questions that should be asked are:

Has the design of the product, in its raw state, been closely examined to ensure that the minimum metal will need to be removed during processing, thus keeping down machining costs and producing the smallest quantity of swarf?

Has the use of chipbreakers and the design of the tooling been considered so that the swarf produced will be in the form most easily and economically handled?

Has the design of the machine tools to be used been considered in relation to taking away the swarf from the vicinity of the cutting tools and ultimately from the machines themselves?

Machining, particularly when modern, high-production machine tools are used, is economical only when the machine tools are maintained at peak production rates, so the first and main objective of any swarf-handling scheme must be to enable continuity of operation to be achieved. In devising a scheme, thought must be given both to economic and other factors, such as:

—installing the system which will give optimum results allied with the lowest initial capital outlay and anticipated maintenance costs;

—choosing a scheme over which effective and efficient control can be constantly and easily maintained;

—ensuring that the best general plant housekeeping is assured;

—being sure that the direct labour concerned will accept and co-operate in obtaining the maximum benefits.

The prime objective of this book is to record information about the effects of all these factors and to make known and discuss the

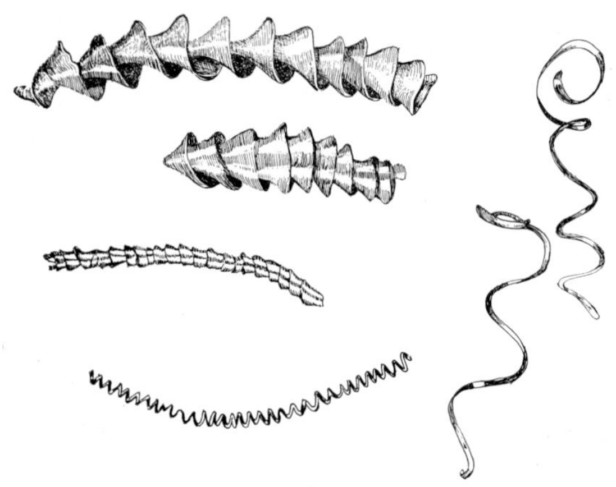

Fig 1.1 Forms of swarf which will accumulate, become interlocked and form bushy masses

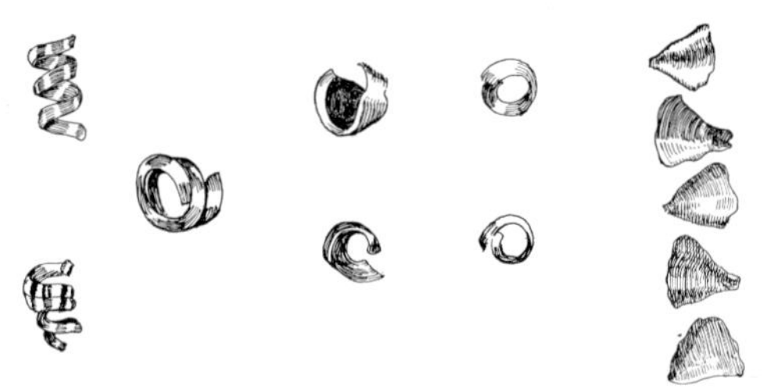

Fig 1.2 Broken chips and short helices which are easily handled and readily conveyed

equipment and methods that are available to assist in solving the problems.

The nature of swarf varies considerably, from particles having dimensions of only a few microns to ribbons of considerable length and cross-section, but to ensure uniformity of terminology throughout the manual, the following classifications will be used (see Figures 1.1 and 1.2):

Swarf that material which is removed in changing the shape of a piece of material, utilising some form of cutting tool employed in a suitable apparatus.

Bushy swarf normally a tangled formation of long helices of the removed material.

Broken chips normally derived by breaking the ribbon of workpiece material at regular intervals. The average length of chip depends on:

workpiece material;

type of cutting tool;

type of chipbreaker.

Dust and sludge normally collections of particles having a diameter less than $\frac{1}{16}$ in (1·6 mm). Dry machining operations produce dust as well as chips of regular length. Dust forms into sludge when cutting fluid is used.

The information contained in the volume has resulted from the expert knowledge and wide experience of senior representatives of about sixty different companies whose interests lie in the fields that have been covered—machine-tool manufacturers, users of machine tools in large and small numbers, makers of mechanical, hydraulic and pneumatic conveying machinery, specialists in filtration and separation, and suppliers of coolants and cutting oils. In spite of this store of knowledge, however, and because of the possible alternatives and the many differences between manufacturing plants, the final answer to a specific problem in any one machine shop will be found only after careful study and analysis of the requirements and site conditions. Furthermore, it is fully realised that the information contained in this edition of the manual will tend to become outdated as time goes on. The manual does not, therefore, attempt to offer solutions to all swarf-removal or handling problems but it will serve to stimulate awareness of the situation, to bring together and publicise much of the available data, and to act as a guide to the whole subject.

2 THE SWARF PROBLEM AS IT AFFECTS THE MACHINE-TOOL MANUFACTURER

INTRODUCTION

Although manufacturers of machine tools must always be aware of the problems associated with the control and removal of swarf from the machines they produce, the machine-tool designer must work with the prime object of producing a machine tool that is an efficient instrument for the cutting of material and the shaping of workpieces. Swarf has been described as the second major product from a machine tool but it is actually the waste-material product from cutting operations and, as such, has always presented a problem in collecting, removing, handling and disposal.

Swarf is produced continuously whenever machining operations are in progress and for a removal system to be completely successful the swarf must be collected as close to the cutting zone as possible and speedily taken away. Swarf that is allowed to accumulate on the machine elements may be responsible for any or all of the following:

impeding operator functions;
causing excessive wear in the machine;
causing inaccuracies in machining;
fouling the tooling, fixtures and workpieces;
menacing the health and safety of the operator;
leading to a low standard of housekeeping.

Amongst the many factors connected with the production of

swarf, two must always receive special consideration because of their influence on the volume of swarf produced. These are:

Component design

This is of great importance as correct design will control the quantity of metal removed as swarf.

Machining conditions

The form taken by the swarf as it leaves the cutting tool is of equal importance and it may be necessary to enlist the aid of chip-breakers or other means to produce the swarf in the most easily handled form.

These factors are more fully dealt with in Chapter 3.

When a customer wishes to order a new machine tool, and particularly if it is to be of advanced or complex design, it is essential that the parties concerned maintain the closest contact at all times. Because the problem of swarf is intimately related to design, only consultation between the customer, the machine-tool manufacturer and the suppliers of suitable swarf-handling equipment during the project and design stages will make possible the incorporation of a satisfactory swarf-removal system that will give economic benefits. It is also important to ensure that sample workpieces provided are exactly similar to those that will be machined in full-scale production.

MACHINING CONDITIONS

(a) The volume of swarf

It is not generally realised how rapidly swarf can become a serious problem by bulk alone, if it is allowed to accumulate.

The volume produced per hour may be calculated in the following manner:

$$V = A \times d \times n \times v$$

where V is the volume occupied by the swarf produced in one hour
A is the area of the machined surface
d is the depth of stock removed
n is the number of workpieces per hour
v is the volume occupied, as swarf, by unit volume of the workpiece material.

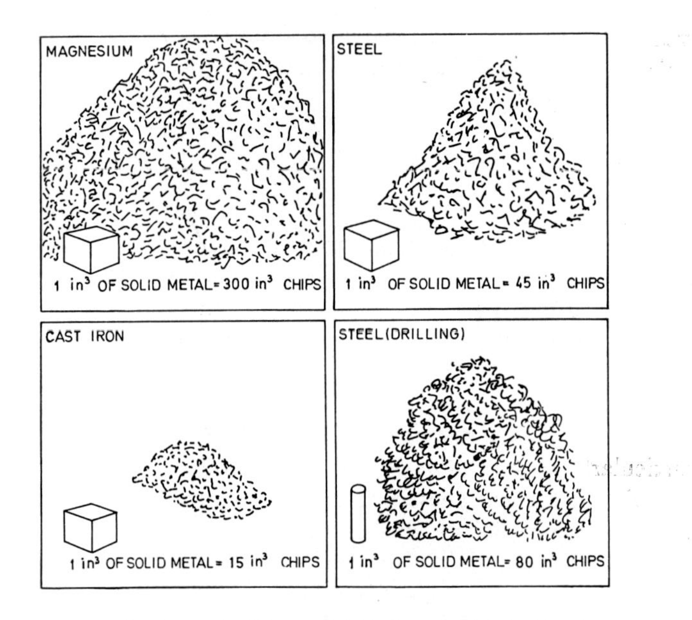

EXAMPLES OF SWARF VOLUMES FROM SPECIFIC MACHINING OPERATIONS.

MATERIAL	TYPE OF MACHINING OPERATION	VOLUME OF SWARF FROM 1 in^3 (16 cm^3) OF SOLID METAL
MAGNESIUM	FACE MILLING	150-300 in^3 (2500-5000 cm^3)
ALUMINIUM	SPAR MILLING	50 in^3 (800 cm^3)
STEEL	FACE MILLING	30-45 in^3 (500-700 cm^3)
CAST IRON	SLAB MILLING	15 in^3 (250 cm^3)
CAST IRON	FACE MILLING	7-10 in^3 (115-160 cm^3)
MEDIUM-CARBON STEEL	DRILLING SIZES 1-3 IN (25-75 mm)	40-80 in^3 (650-1300 cm^3)
CAST IRON	DRILLING SIZES 1-3 IN (25-75 mm)	10 in^3 (160 cm^3)

Fig 2.1 Pictorial examples of swarf volumes

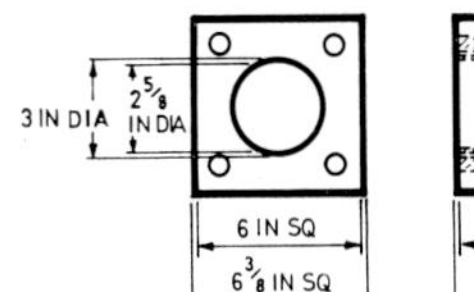

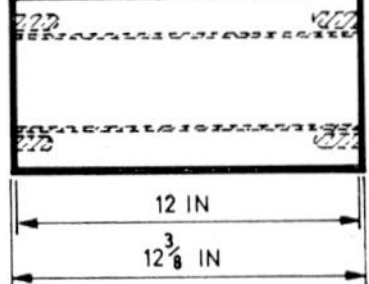

Example A. Cast-iron cylinder

Volume of unmachined casting $=435$ in^3
Volume of machined casting $=340$ in^3
Volume of solid metal removed $=95$ in^3
If 1 in^3 of solid metal $=15$in^3 of swarf chips
Then the total volume of swarf could $=95 \times 15$ in^3
$=$*1425 in^3 of swarf chips.*

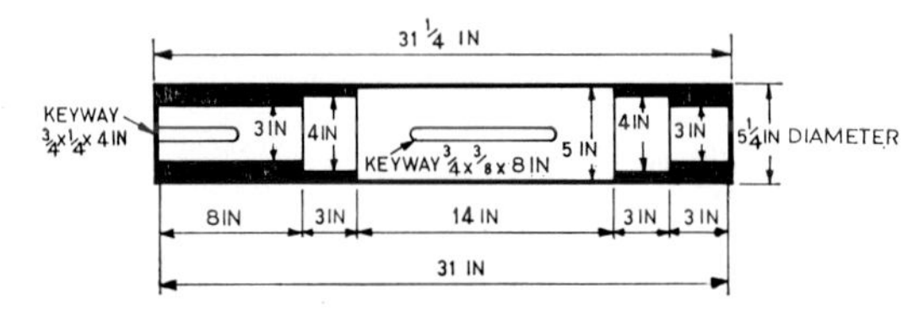

Example B. Low-carbon-steel shaft

Volume of unmachined bar $=676$ in^3
Volume of finished shaft $=425$ in^3
Volume of solid metal removed $=251$ in^3
If 1 in^3 of solid metal can $=45$ in^3 of swarf chips
Then the total volume of swarf could $=251 \times 45$ in^3
$=$*11 295 in^3 of swarf chips.*

Example C. Phosphor-bronze wearing strip

Volume of unmachined material $=220$ in^3
Volume of machined strip $=168$ in^3
Volume of solid metal removed $=52$ in^3
Assuming 1 in^3 of solid metal $=15$ in^3 of swarf chips
Then the total volume of swarf could $=52 \times 15$ in^3
$=$*780 in^3 of swarf chips.*

Fig 2.2 Examples of the volumes of swarf produced during 'every-day' machining operations

Taking the milling of a rectangular face on a grey-iron casting as an example and assuming the following figures:

area of machined surface	$= 12$ in^2 (67 cm^2)
depth of stock removed	$= 0{\cdot}25$ in (6 mm)
number of workpieces per hour	$= 25$
one cubic inch of solid metal	$= 15$ in^3 of swarf (98 cm^3)
then the volume of swarf produced in one hour will be	$= 12 \times 0{\cdot}25 \times 25 \times 15$
	$= 1{,}125$ in^3 of swarf (18,439 cm^3)

Figures 2.1 and 2.2 show the volume of swarf chips resulting from milling and drilling operations on various materials together with examples of the volumes of swarf created during typical, every-day machining operations.

(b) The density of swarf

When considering the removal of swarf from machine tools and the method of transporting it away from the periphery of the machine, it is important to know the density of each type of swarf from each material.

The density, calculated in pounds per cubic foot (kg m^{-3}), depends upon the type of swarf being considered; e.g. in the case of steel swarf produced by a turning operation on a centre lathe, it may be in long spirals or in the form of short chips produced by a tool which incorporates a chipbreaker.

Experimental data on the density of swarf are given in Table 2.1.

(c) Cutting fluids

Although some operations are best performed without the aid of cutting fluid, in general continuing advances in machining techniques, the development of modern materials and improvements in cutting tools necessitate not only the use of the correct cutting fluid but in the most suitable quantities. It is the responsibility of the machine-tool designer to make provision for the cutting-fluid supply and the necessary equipment, and he must rely upon guidance from the customer regarding the materials to be machined

TABLE 2.1 *Experimental data on the bulk density of swarf from different workpiece materials*

	Bulk density lb ft^{-3} (Kg m^{-3})	
Type of swarf	*From published data*	*From actual samples used*
Steel		
Long bulky	15 (240)	32·8 (525)
Short bulky	35 (560)	50·5 (809), 22·5 (360)
Chips	65 (1040)	63·8 (1022), 125 (2000), 107 (1714), 63·5 (1017), 100 (1602), (Mean 92 (1474))
Phosphor bronze		
Light	70 (1121)	96 (1538), 121 (1938)
Heavy	140 (2242)	
Aluminium		
Long	10 (160)	24·4 (390), 22·8 (364), 19·3 (309), 13·7 (219), 3·5 (56), 7·4 (118), (Mean 15 (243))
Short	20 (320)	44.5 (713), 33 (528), 17·5 (280), 22·5 (360), 35·2 (563), 26·7 (427), 21·7 (348), 21·3 (341), 31·8 (509), 14·3 (229), 28·2 (451), 60·0 (961), 15·0 (240), 35·8 (573), 19·6 (314), (Mean 28·7 (459))
Cast iron		
Short bulky		109 (1746), 101 (1618), 89 (1426), 95 (1522), 31·4 (503), 75 (1201), (Mean 83 (1329))
Fine chips		132 (2114), 127 (2034), 94 (1506), 64 (1025), (Mean 104 (1682))

and must certainly know whether the customer intends to make use of cutting fluid.

Consideration must be given to the following points:

1 The designer will need to calculate the capacity of holding tanks required to allow for the volume of fluid concerned. Reliable estimates for the latter are difficult to make as cutting fluid should always feed with high volume and low velocity but the recommended quantities are:

(*a*) For metal-cutting tools the available flow rate should not be less than 5 gallons (22·7 litres) per minute for each cutting tool employed. This quantity may be varied for such machines as would use multi-spindle drilling heads.

(*b*) For grinding machines the flow should be as copious as possible. A minimum of 5 gallons (22·7 litres) per minute or 5 gallons (22·7 litres) per minute per inch (25·4 mm) of face width of wheel, whichever is the greater, is the recommended rate.

Apart from the function of cooling and lubricating the cutting tool and workpiece, cutting fluid available in copious supply offers a relatively simple method of flushing swarf away from the cutting zone towards the point at which it can be removed from the machine.

2 A very important aspect of a cutting-fluid system is the design and placement of the holding tank. If the tank is to be sited within the machine structure, it must be remembered that the temperature of the cutting fluid will rise as a result of its absorption of the heat generated by the cutting operation and the dispersal of this heat throughout the machine elements may cause distortion and lead to inaccuracies in finished workpieces. Cutting fluids may very quickly become degraded by bacteriological action and other contaminants and the tanks should be regularly cleaned; this must be acknowledged by the designer by making the tank of a shape that is easily cleaned and maintained.

3 Care must be exercised in the selection of pumps for use with cutting fluids. They must have the necessary capacity to deal with varying volumes of cutting fluid and be capable of dealing with different viscosities. The siting of the pump is important and care must be taken to avoid the risk of aeration on the suction side, especially when neat oils are being used on grinding machines or gear cutters. Consultation with the pump supplier is recommended in all cases.

The designer will devise the best method of directing the flow of cutting fluid to the cutting zone. The equipment has an important part to play in the functioning of the machine, particularly if the cutting fluid is expected to aid the removal of swarf, and the designer must ensure that the fittings used are of good quality and robust construction; he must ensure that they are positioned so as to give the best service and are rigidly supported; the delivery nozzles should also receive his attention. Far too often the cutting-fluid arrangements on a machine tool appear to have been added as an afterthought and this should not be.

4 It must be remembered that modern cutting fluids with their additives may attack certain materials and the designer must be selective in the choice of materials. In particular, caution must be observed where 'yellow metals', light alloys, zinc alloys and natural rubber are employed.

It is more than likely that the customer will decide upon the final colour he wishes his machine to be but it will be the machine-tool manufacturer who will be responsible for the finishing materials used. Some paints and fillers are liable to attack by cutting fluids, particularly those with certain additives, and to avoid this the designer must make sure that the materials he specifies will withstand such attacks, that the materials are applied in the workshops in the correct manner, and that the recommended drying and curing times are allowed.

(d) Cutting-fluid treatment at the machine

New developments and changes in trends and opinions must always be given early attention by the machine-tool designer and one development in which there is growing interest is the possibility of using continuously cleaned cutting fluids during the cutting operations. Although there are few facts and figures yet available to support the claim, it is seriously contended that the use of cutting fluid from which swarf particles and other contaminants are continually removed will considerably extend the life of the cutting edge of the tool and definitely improve the surface finish of the workpiece.[1]

It is more than likely that customers will begin asking for cleaning equipment to be incorporated in their machines and the designer must prepare by familiarising himself with the equipment available; details of a selection of appliances are given in a later chapter. Briefly, the designer should be aware of the following:

1 Full information on the types of cutting fluids to be used should be made available and the type of equipment agreed with the customer at the outset.

2 If the cleaning is to be by the process of 'settling-out' it will be necessary for the designer to know the average times needed for different materials to settle in the selected cutting fluid, and to allow for the cutting-fluid tanks to be of sizes sufficiently large to accommodate the volume of cutting fluid that will be needed.

3 If the swarf particles are to be removed by magnetic separation then provision will need to be made in the cutting-fluid tank for the magnetic elements and the means of taking away the collected swarf.

4 Should a machining operation warrant the use of a filter which employs a filter medium, then provision will have to be made for the take-off pipes to be of ample size and the cutting-fluid tanks to have a capacity to cover the filtering period and still allow for the desired volume of cutting fluid to be delivered to the cutting zone. It is important to make sure that the flow rate of filtered cutting fluid is high enough.

5 If the customer wishes to use a combined separator and swarf conveyor, early notification should be given to the designer as such appliances can be bulky and will need special design arrangements for their inclusion.

MACHINE TOOL DESIGN

(a) Design to assist swarf removal

There are many features of design that will be incorporated in all machine tools to enable them to fulfil efficiently the functions for which they have been produced but there are several little-used ideas that could assist in the removal of swarf from the cutting zone, if they can be accommodated.

1 On standard machines, particularly so in the case of lathes, vertical slideways can provide a positive solution. Apart from allowing the swarf to fall freely away as soon as it is produced, the

life of the slideways will be prolonged because the swarf will not gather upon them. When vertical slideways are used the structural stiffness of the machine is in no way affected and may well be improved. Such a design could make space available for the inclusion of a conveyor for taking away the swarf, thus overcoming another difficulty.

2 When slideways cannot be vertical, then they should be of the 'herringbone' type and should be so designed that they stand clear of the surrounding structure—see Figure 2.3.

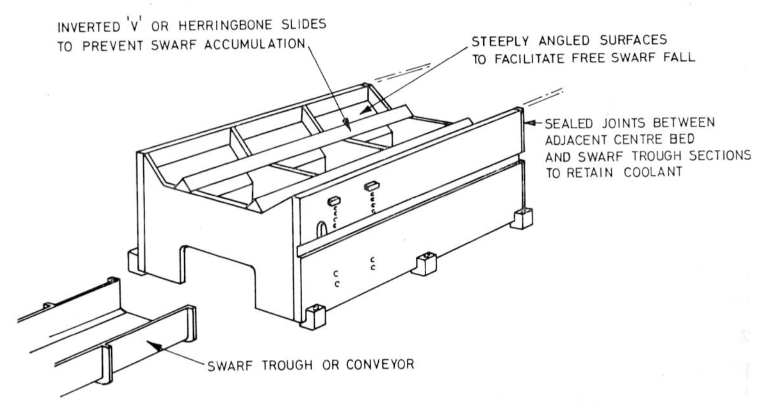

Fig 2.3 A good design of a machine tool bed-section, thought has been given to the elimination of flat surfaces on which swarf can accumulate

3 It is often possible to design machine spindles so that they protrude towards the workpiece through fixed guarding. This arrangement will ensure that the swarf is directed towards the swarf-removal opening and is not allowed to escape towards the machine-head slideways and spindle bearings.

4 The manner in which the cutting-fluid supply is delivered to the cutting zone is worthy of examination. If cutting-fluid feeds can be arranged to pass through the machine spindles then they should be adopted as they will ensure that the swarf is swept backwards and away from the workpiece.

5 Commensurate with structural rigidity, arrangements must be made for gaps of adequate size for the swarf to fall through.

6 When it is known that a machine tool is to be used for 'dry' cutting only, some form of industrial vacuum cleaner could be incorporated into the basic machine.

(b) Minimum slopes down which swarf will freely fall

When cutting fluids are used they increase the tendency for swarf particles to adhere to machine surfaces; it is necessary, therefore, for the designer to know the minimum slopes to prevent such build-up and the minimum flow rates required to flush the swarf away. In an attempt to provide such information, an experimental programme has recently been carried out to examine the following factors: the slope and surface finish of various materials, the behaviour of swarf particles under the influence of different fluids, and the flow rates required with different fluids to move swarf particles along a surface.[2]

Three sets of experiments were carried out to investigate:

1 The behaviour of swarf on dry surfaces, by placing a swarf particle upon a plate and tilting the plate until the particle moved.

2 The behaviour of swarf on a partially wet surface, using a swarf particle placed on a drop of fluid on a plate and tilting the plate until the particle moved.

3 The behaviour of swarf on a fully wetted surface, by depositing a swarf particle on a plate set at a fixed angle and increasing the flow rate of the fluid until the particle moved.

For the purposes of the experiments the following materials were used:

Nine different plates made from cast iron, mild steel and stainless steel, and these had surface finishes that were rough or fine machined, painted or polished.

Two oil emulsions and three cutting oils.

Swarf samples of cast iron, mild steel, phosphor bronze and aluminium.

During the course of the studies, it was found that the behaviour of swarf particles varied widely under apparently identical circumstances.

The initial experiments were with dry surfaces and the results obtained made it possible to compile Table 2.2 which shows the average range and maximum angles.

TABLE 2.2 *Effect of surface finish on the slope of machine tool surfaces down which swarf particles of different materials will roll freely. Angle of slope (degree)*

	Type of finish on plate surface					
Swarf material	*Smooth (2–20 μm CLA)*		*Fine-machined or painted (20–150 μm CLA)*		*Rough-machined (150–300 μm CLA)*	
	average of angles	*maximum angle noted*	*average of angles*	*maximum angle noted*	*average of angles*	*maximum angle noted*
Cast iron	13–26	34	19–30	43	28–45	58
Mild steel	14–60	78	24–45	77	27–72	79
Aluminium	16–46	50	30–32	74	31–72	90
Phosphor bronze	15–24	32	22–25	40		

During the studies it was found that:

1 The slightest contamination by oil on the surfaces increased the angles appreciably.

2 The angle also depends on particle shape: those with relatively high centres of gravity roll easily at low angles but it was not found possible to describe their behaviour in mathematical terms.

3 In many cases relatively high angles were required.

The experiments next carried out used partially wetted surfaces, the conditions studied being those in which a liquid bridge is formed between a swarf particle and a machined surface. An attempt was made to predict bond strengths but so many simplifications had to be made for irregularly shaped particles that the results have little value. However, the following significant factors emerged:

1 With each of the emulsions and cutting oils used, small particles stuck even on surfaces tilted at 90° on all surfaces except some that had been ground.

2 Low liquid surface tension enhances spreading and minimises adhesion but the surface tension of the liquid must be less than that of the solid for spreading to occur. The fact that most swarf particles protrude above the liquid films means that their movement is aided by liquid surface tension.

3 High solid surface tension enhances spreading and adhesion; most metals have high surface energies but are very sensitive to contamination.

4 Low values of liquid-solid surface tension enhance spreading but also increase adhesion. This property is important as it represents molecular attraction across an interface compared with the attraction of molecules for themselves.

5 Surface roughness increases adhesion by increasing both the surface energy and the total area, and it also aids mechanical interlocking.

Thus partly wetted surfaces will be of no assistance in the removal of swarf; however, most machine tools that are meant to be dry are in this condition due to the presence of lubricating oil, etc.

The third phase of this series of experiments was the study of fully wetted surfaces. The flow rates, angles, film thicknesses and fluid velocities were measured and it was found that the thicknesses of film were a little lower and the velocities higher than the theoretical formulae for thin film had suggested. The main experiments were made with an oil emulsion diluted 30:1 and covered the range of plates at slopes of 5°, 15°, 25°, 35° and 45°. The results showed that:

1 At angles above 35°, except on very rough surfaces, all particles were moved by a fluid stream which maintained full flow.

2 The flow rates required were relatively low, e.g. 0·3–2·0 gallons per minute per inch (0·05–0·36 litres per minute per mm) width of surface.

3 The fluid velocities required were between 1 and 3 feet (30·5 and 91·4 cm) per second.

4 In some cases, particles stuck to fully wetted surfaces at higher angles than under dry conditions.

5 It is easier to achieve fully wetted surfaces by using a large number of small jets rather than a small number of large jets.

6 Surface roughness is a major factor and the flow rate required to move the swarf increases with increasing roughness.

Similar studies were made using cutting oils in place of the emulsion. The results obtained were too limited to permit detailed analysis but it was assessed that the flow rates required for cutting oils having viscosities fifty times greater than those of emulsions were generally 50–70% lower.

(c) Wear of machine elements

The sustained accuracy of a machine tool and its length of useful life depend upon maintaining all working parts in a first-class condition and this applies particularly to slideways, sliding shafts and bearings. The continued presence of any kind of swarf on machine slideways is extremely detrimental and will certainly shorten the life of the machine because

1 It will cause excessive wear by abrasion.

2 It will 'blind' oil holes and fill oil grooves, thus restricting the passage of the lubricant and causing the oil film to break down.

3 It may become trapped between two sliding surfaces, causing 'scoring' and probable seizure of the two parts.

Preferably then, swarf falling from the cutting zone should always be directed away from the machine slideways, but as this is not always possible, two other safeguards should be jointly used, these being:

1 Very efficient telescopic metal guards are available, and when

these are fitted and maintained in good condition, they afford the most comprehensive method of protection.

2 Whatever has to move along a slideway, whether it be a table or saddle, must be equipped with wipers at both front and rear. These are efficient so long as the sliding member never leaves the end of the slide.

Increasing interest is being shown in the use of non-metallic inserts of plastic or plastic/fabric for the slideways on the undersides of moving components such as lathe saddles, planing-machine tables, milling-machine tables, etc. There is a number of suitable materials available for this purpose, the better-known ones being made up from asbestos fabric sheets, impregnated with phenolic resin, that are assembled in layers and compressed to a predetermined density between heated platens. The resulting material can be machined in a manner similar to cast iron, it may be hand-scraped and ground, and it may be attached to a suitably prepared surface by an epoxy-resin adhesive. The use of these materials under heavy operating conditions offers a number of advantages:

1 Wear of the metallic slideway of the machine bed is minimised.

2 Swarf particles that are trapped between the sliding surfaces become embedded in the non-metallic material and scoring of the mating surface is prevented.

3 These laminates are unaffected by oil and copious flows of standard lubricants may be used in the normal way to flush the swarf from the slideways.

4 When the inserts are worn and require replacing, this is a simple matter and the new inserts may be quickly bedded to the unmarked and polished metallic slideway.

It is equally important to protect the surfaces of machine spindles and sliding or rotating shafts. Whenever swarf is produced, a certain amount will be in the form of particles that are only a few microns in size and, if these are allowed to penetrate to the bearings, the life of the latter will be drastically reduced. The answer is to fit guards or deflectors capable of directing the bulk of the swarf away from such moving parts and towards the swarf track, to place efficient wipers immediately adjacent to the location at which the shaft enters the headstock or other unit, and to consider the use of some form of standard seal. If two seals were fitted 'back-to-back' they would serve a double purpose by retaining oil within the unit and restricting the entry of swarf.

Regular cleaning of a machine tool is always to be recommended. However, when air jets are used for blowing away the swarf and cutting fluid from location points, care must be taken to ensure that swarf is not forced past seals and wipers.

(d) Lubrication

Whilst the adequate lubrication of machine tools is recognised as necessary to their successful working, inefficient methods of lubricating can complicate the removal of swarf under both wet and dry cutting conditions. With the extending use of hydraulics as a driving medium, the possibility of the hydraulic fluid entering the cutting fluid must be recognised and guarded against. In acknowledging the hazards, the attention of the designer should be directed towards the following:

1 To guard against any contamination of the cutting fluid by incompatible oils, efficient sealing arrangements are essential. It must be remembered that such contamination will not only hinder swarf removal but will also change the nature of the cutting fluid and could neutralise the effects of the additives.

2 Lubrication fittings must be of good quality and strongly made; oil pipes made from thin-walled tubing can be quickly damaged when in regular contact with swarf.

3 The positions of pipes for supplying cutting fluids, lubricating oil and hydraulic fluid should be carefully planned so that they are not easily damaged, they are freely accessible for maintenance and they do not provide catchment areas for swarf. Where possible, pipes should be run under ledges but when this is not possible they should be covered by guards having a sloping upper surface.

4 When the swarf-removal system is designed so that the cutting

fluid will be used to clear the cutting zone and direct the swarf towards the conveyor, the angles of various parts of the machine structure in contact with swarf will have been determined so that the flushing action of the cutting fluid will keep them clear. Should any oil of a greater viscosity become mingled with the cutting fluid then the angles will probably be insufficient and the swarf will adhere to the surfaces.

5 Slideways, with other parts of a machine tool, will need to be lubricated. They should not be subjected to the flushing action of the cutting fluid, resulting in the removal of the lubricant as well as the swarf. This is dangerous and both guarding of the slideway and deflection of the cutting-fluid flow are recommended.

6 Under dry cutting conditions, it is inevitable that a certain amount of fine metallic dust is created. Wherever oil is present on machine surfaces this dust will be trapped and held, rendering complete swarf removal impossible and adding materially to the difficulties of good housekeeping.

(e) The design of tooling

With the growth in the use of automated transfer lines and multi-station machine tools, the design of tooling becomes ever more important and has a distinct contribution to make towards the efficient removal of swarf. The following basic design features are worthy of consideration in all cases.

1 Always avoid overcrowding in the tooling area.

2 Whenever possible, shrouds or deflectors should be arranged around or near the cutting tools so as to direct the swarf towards prepared take-off points.

3 Location surfaces on jigs and fixtures should be minimal in area and easily cleaned.

4 No surface in the line of swarf or cutting-fluid fall should encourage the build-up of cuttings; even locating pins should be shaped—see Figure 2.4.

5 Locations for components must be free from swarf traps and

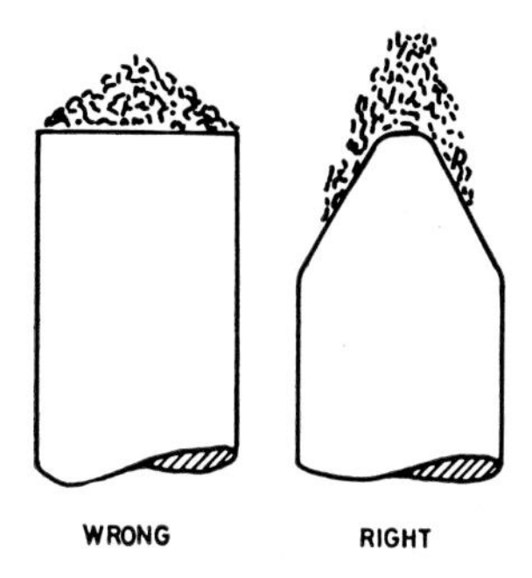

Fig 2.4 The illustration clearly shows how swarf is encouraged to fall away from the top of the cone-pointed locating pin

holding fixtures should be without pockets or grooves. If it is necessary for a component to be located in a bush then care must be taken to ensure that swarf can fall away without obstruction—see Figures 2.5 and 2.6.

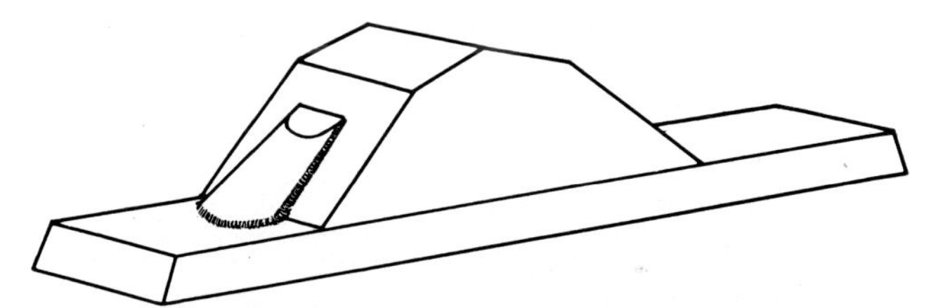

Fig 2.5 A drawing of a jig component—note the angled surfaces to prevent swarf collection

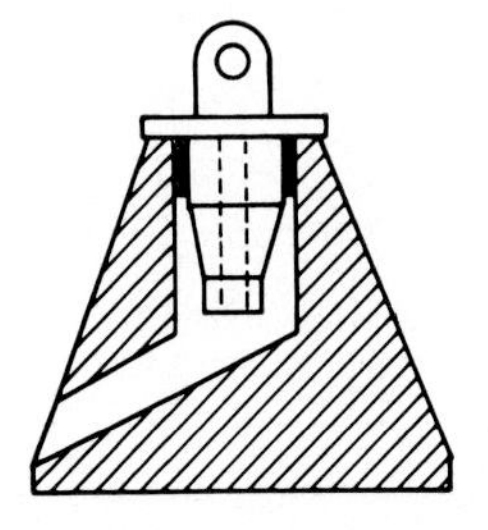

Fig 2.6 A milling jig that in all respects will aid the clearance of swarf

6 Locators and plungers should be raised to clear any accumulation of swarf and location bushes should be positioned so that swarf will not accumulate inside—see Figure 2.7.

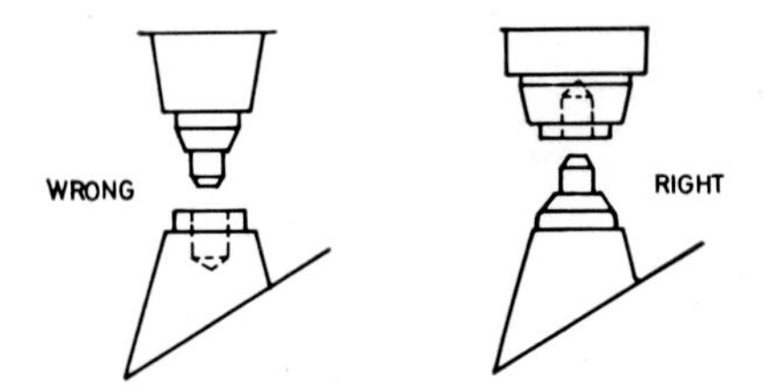

Fig 2.7 The right and wrong of location design

7 The design and location of plungers is important. They should be positioned so that there is no possibility of swarf collecting between locating surfaces and they should not be so long as to allow swarf to accumulate behind them—see Figure 2.8.

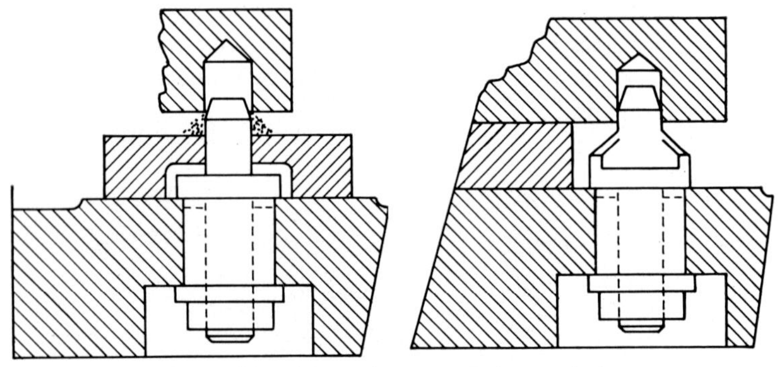

Fig 2.8 The right and wrong of plunger design

8 Whenever it is necessary to use bush plates for drilling operations, the shape of the plate should be selected with care; Figure 2.9 shows how such plates may provide catchment areas for swarf.

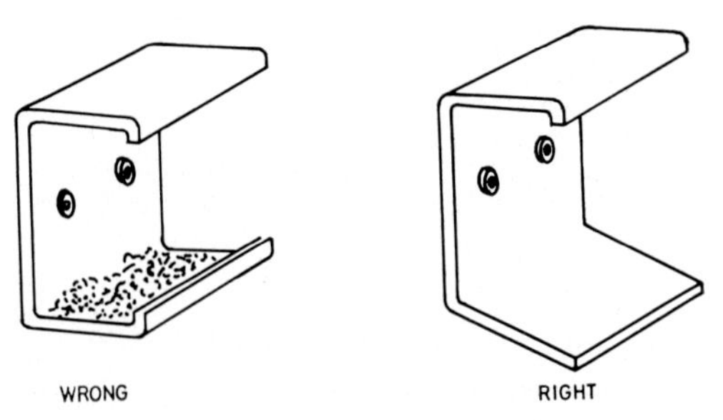

Fig 2.9 The right and wrong of drill jig design

9 Good practice in fixture design and the use of swarf deflectors for the protection of slideways clamping and plunger mechanisms is shown in Figures 2.10 and 2.11.

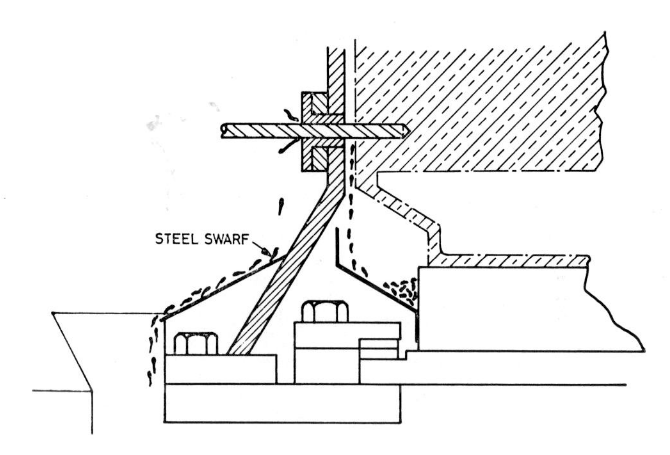

Figs 2.10 and 2.11 These two illustrations show the use of deflectors in guiding the swarf away from machine-tool surfaces

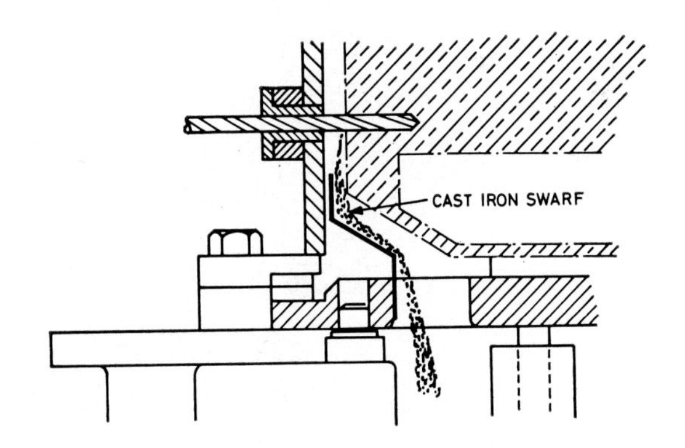

10 A fundamental requirement of jig-and-fixture design is that flat surfaces should be eliminated wherever possible and such tooling, together with platens and tables, should have inverted 'V' or cone-type cross-sections with smoothly finished surfaces to make it easy for swarf to fall away.

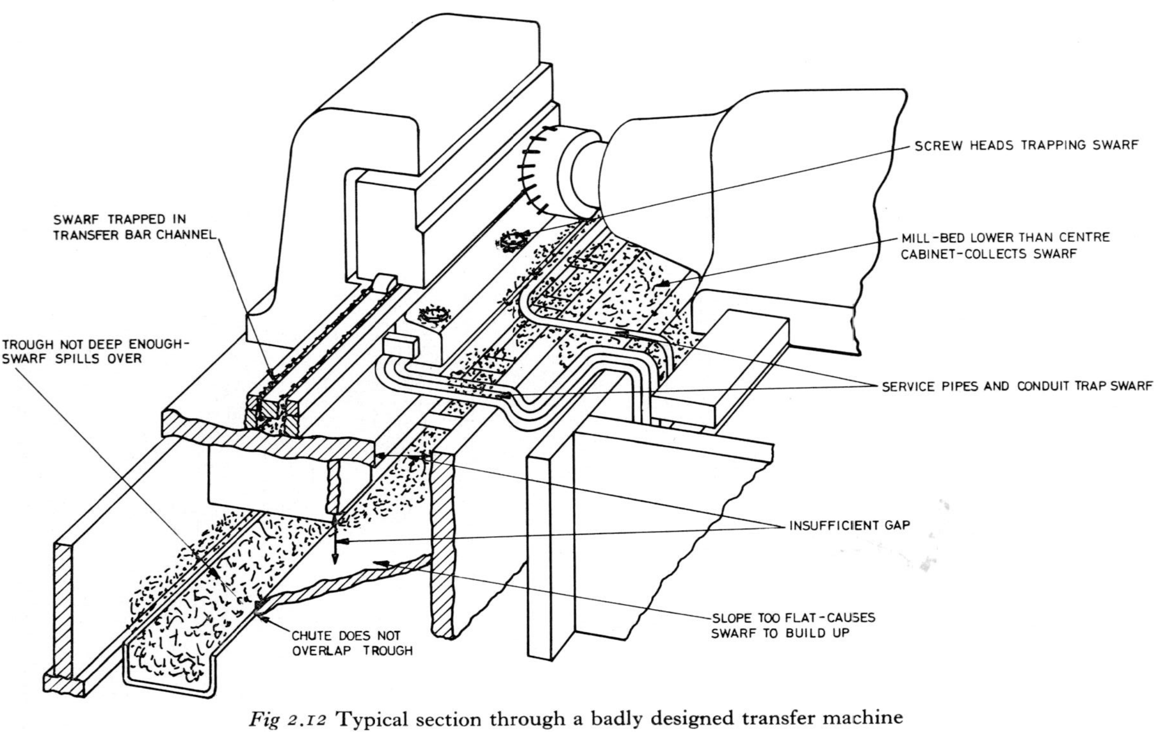

Fig 2.12 Typical section through a badly designed transfer machine

Because of the nature of transfer machines, their complexity and the fact that the machining of the workpiece must be carried out with the minimum of manual assistance, the design of the machine structure must include correctly angled surfaces to aid the fall of swarf, adequate gaps through which the swarf can fall, correctly sited and aimed cutting-fluid delivery pipes and nozzles to clear the tooling, and the necessary space along the length of

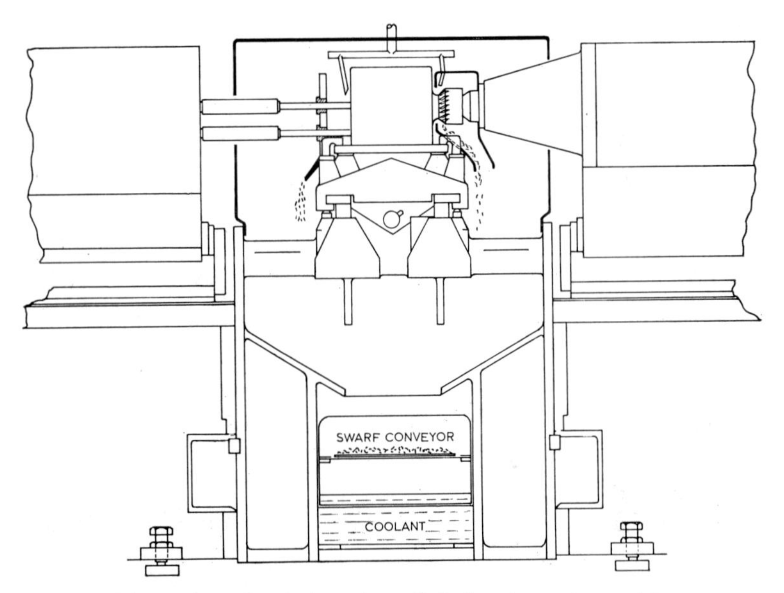

Fig 2.13 A sectional view of a well-designed transfer machine

the machine to accommodate a suitable swarf conveyor. Figures 2.12 and 2.13 illustrate what may be considered bad and good design in transfer machines for the control and removal of swarf.

(f) Guards

Having cutting fluid flowing at the volumes recommended, it is very necessary for the machine-tool designer to pay particular attention to the question of providing adequate guarding. The purpose of guards is initially twofold, they must afford complete protection to the operator and must be so designed that they will retain all swarf and cutting fluid within the periphery of the machine; if the cutting fluid is to assist in the removal of the swarf then the guards should be designed and positioned so that they direct both swarf and cutting fluid towards any in-built removal equipment.

There are a number of recommendations that should be heeded when considering the basic design of guards.

1 Guards should be as robust as possible and rigidly supported where this can be done.

2 Guards which are not permanently fixed should be designed so that they are easily removed when necessary and are just as easy to replace.

3 Guards designed on the overlapping, drip-proof principle have advantages over the sealed, abutting type.

4 Any brackets or framework used in fixing the guards should be so placed that they do not become a collecting place for swarf.

5 The cutting of apertures in guards to allow for the passage of pipes, conduit or other machine elements should be avoided. Such arrangements tend to destroy the purpose for which the guard is needed.

6 When large volumes of aluminium swarf in chip form are being produced guards must provide 100% cover. This type of swarf will enter any small opening and, if a cutting fluid is used, the quantity of swarf that avoids collection is alarming.

7 Pipes, conduits, brackets, etc., fitted to external structure surfaces may all form swarf traps and whenever possible pipes should be routed inside castings under flanges or, if such fittings must be placed where they can cause obstructions, then they should be protected by easily removed guards or covers.

8 Because of the scatter of swarf chips, milling machines are difficult to guard, particularly when dry cutting. The designer could examine the methods used on some automatic lathes for total enclosure, such guarding being very effective in controlling both swarf and cutting fluid as shown for example in Figure 2.14.

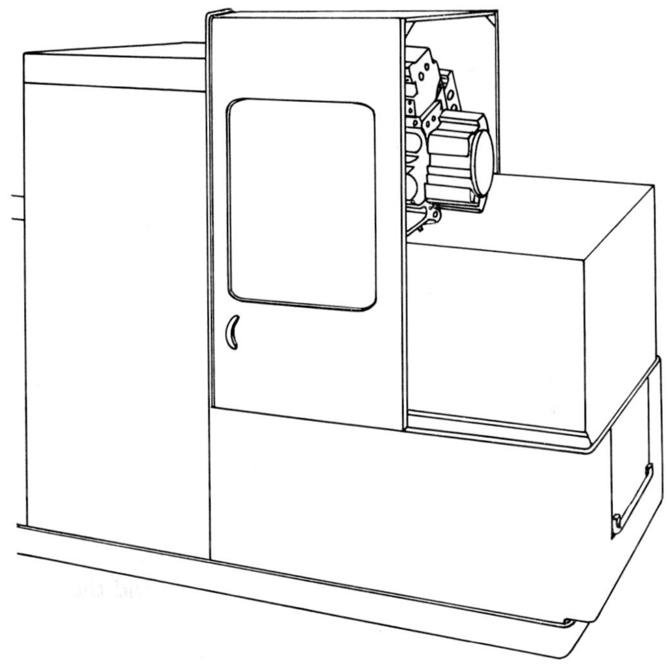

Fig 2.14 An example of efficient guarding applied to a single-spindle automatic chucking machine. The illustration shows the front and rear sliding guards and chip guard in position giving total enclosure for swarf and coolant and offering maximum protection for the operator and full use of coolant at optimum speeds

SWARF-REMOVAL SYSTEMS

In the preceding sections, stress has been laid on the need to remove swarf from the cutting zone but, having achieved this, the next step is to take the swarf away from the machines. At times, it is not possible for a customer to give much thought to the need to remove swarf from his new machine tool but it is he who will need to be satisfied with the designer's efforts and he should be made aware of the situation from the outset.

The machine-tool manufacturer is always conscious of the economic benefits that can be derived from a successfully operated swarf-handling system but, too often, customers show little interest in organised systems because of the capital costs involved. Such costs can be rapidly recovered where an efficient system is installed, some of the factors involved being:

1 Swarf is the enemy of machine-tool slideways, bearings, lead-screws and nuts, elevating screws and nuts, etc. Efficient control of the swarf will reduce the expenditure on replacement parts and reduce maintenance costs.

2 Cutting fluid that is retained within the machine or system is availablc for rc-usc but cutting fluid sprcad around the workshop floor is lost and must be replaced at a cost.

3 Uncontrolled swarf can damage the finished surfaces of work-pieces, necessitating salvage operations, or it can be responsible for total ıejection and consequent loss of material, labour and overhead recovery, to say nothing of loss of profit.

4 Individual cutting-fluid tanks require regular attention in the way of cleaning and sterilising, and cutting-fluid pumps are subject to quite rapid wear. With an organised system or central system such maintenance costs are considerably reduced and a positive economy is made.

5 Cutting oils are expensive and should be separated from the swarf for re-use; the savings can be impressive.

6 Bushy swarf creates its own problem for the scrap-metal merchant. Broken or briquetted swarf will command a significantly higher price.

(a) Swarf removal from individual machines

Single machine tools, whether of the multi-station type or not, present their own swarf-removal problems and these can be quite acute in the case of grinding machines where a high degree of accuracy and surface finish is required.

The actual removal of the swarf from the machine will, in many cases, still be most economically done by manual methods but the designer should know that there are conveyors available that can be set into the machine base to catch the swarf, elevate it and deliver it to a suitable bin for disposal. A typical example is illustrated in Chapter 4 and such a conveyor can be used with capstan, turret and centre lathes and should not necessitate alterations to the machine structure.

With a multi-station machine, some form of built-in conveyor will almost certainly be necessary and, if ductile materials are being processed, then chipbreaker tools should be used to avoid the machine becoming choked with 'bushy' swarf.

Modern machine tools and cutting tools are capable of removing stock at a high rate; in doing so, there is likely to be a significant increase in the heat generated both by the moving parts of the machine and the swarf and this can affect the accuracy of the machine. This is a problem for the designer and the most effective answer is a copious supply of cutting fluid to wash away the swarf from the machine structure and to dissipate the heat.

The swarf produced in grinding operations will mostly be metallic but will be mingled with grinding-wheel debris in the form of grit and bonding material and, because of the cutting fluid, will be in the form of 'sludge'. To remove such swarf the machine designer has several courses open to him, such as:

1 The cutting-fluid tank can be equipped with a magnetic separator that should collect the majority of the metallic (ferrous) particles but will not arrest the grinding-wheel debris.

2 The cutting-fluid tank can contain a series of weirs to deal with any floating contaminant.

3 The tank can be made sufficiently large for it to act as a settling chamber and the contaminants may settle to the bottom.

4 The design can be such that an efficient filter is incorporated, capable of arresting 90% of all solid particles down to a size of 2–5 microns. Cutting fluid that has the swarf content removed to this extent will prove beneficial in extending the life of the grinding wheel between dressings and will show an improvement in surface finish.

(b) Central systems

With the advent of transfer machines and their increasing use, resulting in the production of large volumes of swarf, it is essential that they are equipped with an efficient swarf-removal system coupled with an effective arrangement for the retention of cutting fluid, if this is used. To a somewhat lesser extent, similar remarks may be applied to groups of single machines operating on a line basis and producing the same type of swarf.

When transfer machines are used, it is likely that several will be sited in close proximity to one another and such a layout can be efficiently served by a central system for dealing with the swarf, ensuring the cutting-fluid supply and cleaning the cutting fluid, and this simplifies the task of maintaining the cutting-fluid mixture in the most suitable proportions. A central system will usually be a combination of mechanical and hydraulic conveying when cutting fluid is used, and mechanical and possibly pneumatic conveying when cutting conditions are dry.

The machine-tool designer should note that central systems offer the following advantages:

1 The method of swarf collection and cutting-fluid retention coupled with fewer swarf-discharge stations provides for lower-cost housekeeping and improved working conditions.

2 The use of central systems has promoted rationalisation of cutting fluids and reduced the variety that tended to be used with individual machine arrangements.

3 Generally, the floor-space requirements are reduced when 'in-floor' central systems are used. Without doubt, the problem of swarf removal from large, special transfer machines is most effectively solved by the use of trenches in the floor. Unfortunately, there appears to be an increasing resistance to the use of trenches by some customers.

4 Maintenance costs are reduced since the need to service a multitude of cutting-fluid pumps, filters and cutting-fluid tanks is eliminated.

5 There is a greater opportunity to control the bacteriological contamination of the cutting fluid.

Naturally, central systems have certain disadvantages but it is not considered that these outweigh the advantages. They are:

1 Central systems fix the positions of machine lines and hence the implications of any future rearrangements must always be considered at the outset.

2 A breakdown in the system can result in production being lost from a complete transfer line or group of machines.

3 It is necessary to guard them securely and to maintain strict control as they can become convenient disposal areas for rubbish.

An illustration of a central system serving transfer lines is given in Figure 2.15.

Central systems may be used with advantage for groups of similar or dissimilar machines where the coolant is suitable for all the operations and the materials are such that the swarf does not require segregating. Again, a main drawback may be that any reorganisation or the addition of further machine tools can necessitate new trenches, the discarding and filling in of some existing trenches, and the rearrangement of cutting-fluid tanks, cleaning apparatus and conveying equipment.

SWARF CONVEYING

It is possible that many machine tools will not require special swarf-handling facilities as the rate of swarf production may be such that it can be handled easily by the operator or by a labourer. However, there are many instances when single- or multi-cutting heads may be grouped together either in the form of a special-purpose machine or as a result of forming a bank of similar machines. Under these circumstances the volume of swarf produced may demand the use of proper mechanical handling facilities. Depending on the nature of the machining operations, type of machines, the use of cutting fluid, etc., it is possible to remove the main bulk of the swarf by one or a combination of different types of conveying system:

(a) Mechanical conveying

1 If mechanical conveyors are to be used, the base of the machine will need to be specially designed to accommodate them

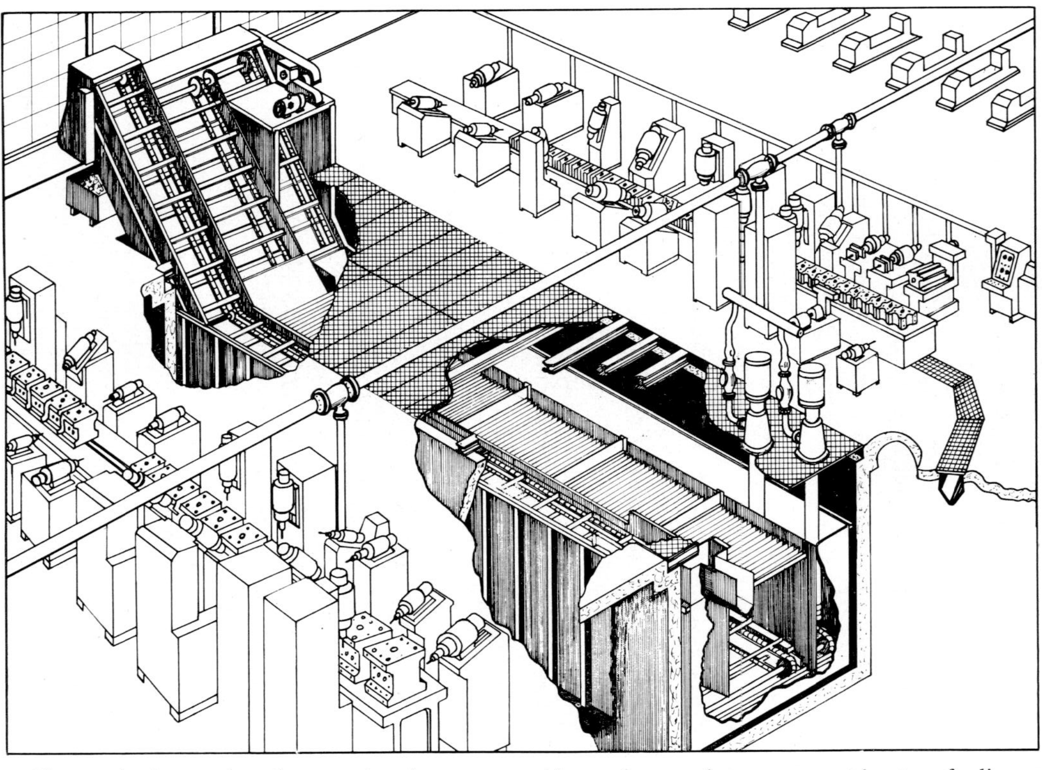

Fig 2.15 An impression of a central coolant system with swarf-removal conveyors serving transfer lines

if they are not to cause unacceptable reductions in machine rigidity. Also, conveyor vibration may affect the accuracy of the machine tool if it is not properly isolated.

2 Selecting a suitable size of conveyor can be difficult unless a clear indication can be given of the amount of swarf anticipated.

3 Because most machine tools conform approximately to a certain shape related to type, before the designer can arrange for the inclusion of a conveyor the customer must state his 'take-off' requirements as these may control the positioning of the equipment.

4 A mechanical conveyor has many moving parts which will be subject to wear and so access for servicing must be easy.

5 Many machine tools have the cutting-fluid tank contained within the structure or require it to be positioned very close by; in either case this could create a problem for the designer as little room would be left for the inclusion of a conveyor.

(b) Hydraulic conveying

1 As different materials behave differently under the influence of hydraulic power, the designer will need to know the type of swarf that will be produced so that suitable angles of slope can be provided on relevant surfaces.

2 For hydraulic conveying to be fully efficient, there must be no obstructions that can hinder free flow. This means that the tracks or channels should be designed without discontinuities or sharp bends.

3 The size of the cutting-fluid tanks will be controlled by the volume of the cutting fluid required for the transportation and the distance the swarf is to be conveyed. This information is essential to the designer as he must arrange for an adequate flow of cutting fluid to be delivered to the cutting zone at all times.

4 Flow rates vary with the viscosity of the conveying fluid and this can affect the angles of slope of swarf tracks and channels (see Table 2.2). For instance, the flow rate required for transport-

ing swarf in a cutting oil with a viscosity fifty times that of an emulsified oil can be 50–70% lower. Therefore, the type of the cutting fluid to be used must be known to the designer in order that suitable arrangements can be made.

When cutting fluid is to be the medium for removing swarf from the cutting zone and transporting it to a suitable collecting point, the latter poses an entirely different problem to the machine-tool designer, particularly on transfer machines.

The first part of the problem is concerned with the flow of a liquid in an open channel and, although much work has been done to establish data on liquids in pipes, little relevant information is available on open channels.[3] However, some work has recently been carried out using a rectangular trough 18 in (457 mm) wide and Table 2.3 summarises results which emphasise the importance of slope and the relatively small effect of volume on velocity and hence on the size of swarf particle transported.[2]

TABLE 2.3 *The effect of slope and rate-of-flow on velocity of travelling fluid in a velocity trench*

Flow	*200 g.p.m. (9092 l min⁻¹)*		*400 g.p.m. (18184 l min⁻¹)*	
Slope	*Velocity feet min^{-1} (m min^{-1})*	*Depth inches (mm)*	*Velocity feet min^{-1} (m min^{-1})*	*Depth inches (mm)*
1 in 12	375 (114·3)	1·0 (25·4)	490 (149·35)	1·55 (39·4)
1 in 48	240 (73·15)	1·45 (36·7)	310 (94·49)	2·30 (58·4)
Level	105 (32·00)	3·2 (81·2)	145 (44·20)	4·7 (119·4)

Because of the wide variation in the shape and size of swarf particles it is not surprising that, when immersed in a fast-moving stream of cutting fluid, their dynamic characteristics do not follow any particular law. Thus, it is extremely difficult to simulate the behaviour of swarf on a scientific basis and to predict how a particle will behave under the conditions imposed by hydraulic methods of transportation. Nevertheless, in addition to the results shown in Table 2.3 it is possible to make general statements which may be summarised as follows:

1 The drag forces on irregularly shaped swarf particles are very much higher than those on spheres of equivalent weight. Consequently, the terminal settling velocities are between 15 and 88% of those for spheres, most lying between 22 and 45%.

2 The effect of size, density and shape of particles presented difficulties in developing relationships but the results obtained for the particles examined showed that velocities were generally between 2 ft (600 mm) and 3 ft (914 mm) per second but velocities of 7 ft (2134 mm) per second were required in some cases.

3 The slope of the channel is a major factor because of its effect on liquid velocity. A typical swarf sample dropped into the trough when level required 100 g.p.m. (454·6 $l\ min^{-1}$) for transport but only 33 and 3·1 g.p.m. (150 and 14·09 $l\ min^{-1}$) with slopes of 1 in 48 and 1 in 12 respectively.

4 Experiments with dropped versus stationary particles showed that the medium and maximum velocities required were mostly 20–30% higher for stationary particles.

5 The addition of jets can create an artificially high velocity in the channel.

The second part of the problem concerns the capacity of the cutting-fluid system required for a given installation, and experience gained over the years must be called upon to assist in finding the solution. Because of variations in operating conditions, it is difficult to make any definite recommendations but consideration of a typical installation may provide a suitable example of the factors making up the problem.

The equipment concerned comprised one 35-station transfer machine supported by two small machines for ancillary operations and two stations for washing platens and components. The production rate was one component every two minutes and the estimated production of swarf was 5 lb (2·27 kg) per minute. The tank for the settling-out of large particles was estimated to require a capacity equal to the volume for 6 minutes running time.

Estimates of the cutting-fluid flow rates were made as follows:

Cutting stations	1240 g.p.m.	(5637 l min^{-1})
Washing stations	100 g.p.m.	(454 l min^{-1})
Swarf flushing	500 g.p.m.	(2273 l min^{-1})
TOTAL	1840 g.p.m.	(8364 l min^{-1})

Approaches were made to a number of manufacturers for the purpose of obtaining information to support the above estimate but the recommended flow rates varied widely as did the estimates of the treatment tanks for the cleaning of the cutting fluid.

Such variations will occur in estimated volumes between different manufacturers and this emphasises the necessity for a design office to tabulate the very fullest details gleaned from each installation that is successfully engineered and successfully operated. This is the surest way in which to obtain the facts and figures on which future estimates of cutting-fluid requirements can be based. Transfer machines have been in use for many years and there must be a large accumulation of relevant knowledge available on which the designer can draw.

(c) Pneumatic conveying

1 These systems are most effective when swarf from dry cutting has to be removed. Swarf containing a small amount of cutting fluid can be handled but it is not recommended.

2 As pneumatic systems are designed to collect the swarf as it is produced at the cutting tool, close-proximity hooding may be called for and allowance must be made for this at the cutting zone.

3 A pneumatic system usually requires trunking for the retention of the air and to form the conveyor. The designer must bear this in mind because he must allow for the space such trunking will occupy.

4 If the system is to serve only a single machine or small group of machines, then the designer must know the take-off positions and, in the layout, must make space available for the fan and the collecting unit.

5 It is again important for the designer to know the approximate

volume of swarf that will be produced so that the power of the extraction fan can be correctly calculated.

A different method of removing swarf by pneumatic means has recently been developed for an installation in the U.S.A. where dry cutting is normal on cast-iron components. In the past, the cast-iron dust caused by machining has created a problem in spite of the collector ducts around the machine head and spindles; this method operated against the forces of gravity and the ducts became thickly coated with oil and dust.

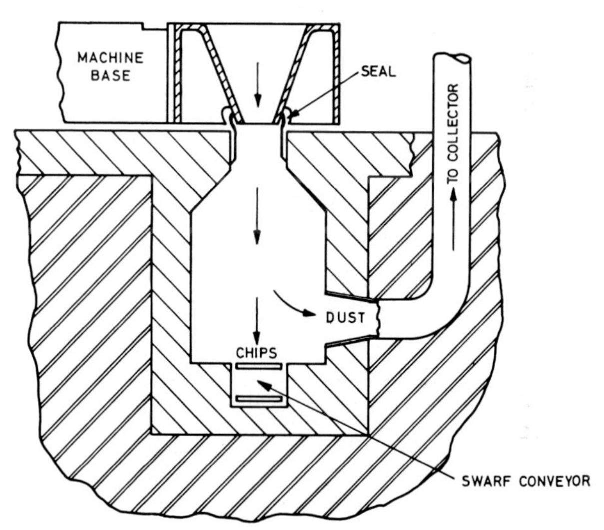

Fig 2.16 A sectional view of a pneumatic extraction system where gravity assists the clearance of swarf from the cutting zone

The changed method is such that the gravitational fall of the swarf is assisted by down-draught ventilation. This arrangement requires a trench, approximately 8 ft (2438 mm) deep, extending the full length and under the centre of the machine line; dust collectors are located at intervals alongside and are connected to the trench by 26 in (660 mm) diameter tubes. The bottom of the trench is occupied by a swarf conveyor. The machine bases are designed to give a 12 in (300 mm) square opening to the trench and to allow a smooth flow of air; the openings are sealed to the trench and the calculated air flow is 2,500 cubic inches (40,975 cm^3) per minute. Because the machine-base openings are sealed, a uniform vacuum is maintained for the full length of the trench and leakage of air from other areas is avoided. Figure 2.16 shows a cross-section of the arrangement.

THE MACHINE-TOOL DESIGNER AND TECHNICAL INFORMATION

When called upon to create an efficient swarf-removal system, the designer may be handicapped in some respects by a lack of general information and a certain scarcity of technical data.

1 There appears to be a scarcity of helpful literature.

2 There is a lack of standardisation in certain fields and this can mean that the designer is not aware of the sizes of equipment for which he must make space available.

3 There has been a shortage of reliable information on some types of ancillary equipment. The final design may be in the nature of a compromise that is arrived at by the process of trial and error.

4 There has been little accurate knowledge available which describes the manner in which swarf particles will cling to surfaces or of the physical laws governing their behaviour. Recent research work has gone some way towards relieving this situation; some details of this work appear in the section of this chapter devoted to the effects of cutting fluids.[2]

5 Fundamental knowledge describing the behaviour of solids when they are being hydraulically transported in open channels is scarce although recent research work has established some factors. The information now becoming available will provide the designer with such parameters as the minimum velocity required, minimum recommended slopes for channels, volumes of liquids required to maintain swarf in suspension and move it along, etc.

6 Valuable information would be more facts and figures specifying the degree of cleanliness to be maintained in cutting fluids in order to achieve the longest possible life for the tool cutting edge and the best surface finish on the workpiece.

These are some of the fields in which the designer believes further research work could be of inestimable value to him, simplifying his task and assuring the customer of an efficient piece of equipment complete with an effective swarf-removal system. Nevertheless, by working closely with equipment suppliers, a designer should be able to obtain specific data and performance guarantees to support the knowledge he has gained over the years.

REFERENCES

1. 'Longer tool life from clean coolant'. *Metalworking Production,* Volume 109, number 5, 15th December 1965, pp. 87–9.
2. 'Research on swarf-removal systems', I. C. Schomburgk. *Machinery (London)*, Volume III, number 2,857, 16th August 1967, pp. 313–21.
3. 'Transporting solid materials in pipelines', E. Condolios and E. E. Chapus. *Chemical Engineering,* Volume 70, numbers 13, 14 and 15; 24th June, 8th July and 22nd July 1963; pp. 93–8, 131–8, 145–50.

3 THE MACHINE-TOOL USER'S APPROACH TO THE REMOVAL OF SWARF FROM MACHINE TOOLS

INTRODUCTION

When setting out to establish the most suitable method of dealing with the swarf produced, it must be realised that the neglect of or mishandling of swarf may bring other problems into being, not the least being the following:

1 Swarf can cause injury to operators due to the sharp edges and the retained heat.

2 Swarf can damage machine tools, the parts particularly exposed being tables, slideways, leadscrews and bearings.

3 Swarf will damage the machined surfaces of workpieces, and swarf that becomes trapped in drilled holes can interfere with subsequent operations.

4 Swarf that accumulates on jigs and fixtures can interfere with the correct positioning of workpieces and lead to inferior quality or scrap.

5 Housekeeping costs are involved in machine cleaning and swarf removal.

6 If the swarf-removal facilities are inadequate productive capacities will not be fully utilised and time will be lost.

The high productivity made possible by machine tools capable of high rates of metal removal has created a need to improve existing methods and to develop new methods of dealing with the increased quantities of material removed from workpieces in the form of swarf. The problem of dealing with large quantities of swarf is aggravated by the fact that high-production-rate machines require the use of copious supplies of cutting fluid and, although the cutting fluid tends to flush the swarf away from the cutting zone, both swarf and cutting fluids must be removed from the machine for separating so that the cutting fluid may be cleaned prior to being re-circulated.

The MUST objective of a swarf-handling scheme is that the swarf be continuously removed from the machine to allow continuity of operation. Also, when cutting fluid is used, consideration must be given to removing and returning it to the machine tool with an acceptable standard of cleanliness. All other requirements connected with a swarf-handling scheme are WANTS.

But swarf—see Figure 3.1—from sophisticated and expensive machine tools, is costly to produce and, to maximise revenue from the sale of the swarf, it is desirable to minimise the over-all cost of handling it between the point of generation and the point of sale. The minimising of costs must also be consistent with:

1 the ability to maintain effective and efficient control over the whole operation;

2 the maintenance of good plant housekeeping;

3 ensuring a dignified handling system for employees.

In general, the efficiency of handling swarf has failed to match the accelerating volume of production and the application of new production techniques. The handling of swarf presents problems that require careful analysis to ensure effective and economic solutions; as labour costs continue to rise, cost reductions from the efficient handling of swarf must be continuously sought. A co-ordinated swarf-handling system must be regarded as an integral part of the manufacturing process; the application of new methods to fragmented parts of a factory may only prove to be expedients. A complete appraisal is required of the whole pattern of swarf production before choosing the method and equipment.

Sophisticated mechanisation is not necessarily the answer to all swarf problems; indeed, the ill-considered adoption of mechanisation may only add to overhead costs and do nothing to improve efficiency. Some companies may view a capital investment in swarf-handling equipment as an unnecessary asset acquisition but this view is valid only if the operating cost of the current method is known and has been compared with that of a new method.

Fig 3.1 The photograph clearly shows the conditions arising from lack of control of swarf and coolant. Note the dangerous condition of the floor and the generally bad housekeeping

The final selection of a swarf-handling system must depend on a number of factors that will have different impacts and solutions for each type of plant. The best way of handling swarf in one plant is not necessarily suitable for dealing with a similar problem in another plant and many alternative solutions can apply between the extreme limits of a fully automatic or a fully manual approach. The contents of this chapter can only provide guidance to the analysis process and cannot specify packaged solutions. The right system can be defined only as the one that is appropriate to the nature of the problem, the selection criteria being minimum capital expenditure and operating cost coupled with efficient swarf handling.

To date, the effective resolution of swarf problems has been hampered by:

1 split responsibility and lack of communication between machine-tool designer, swarf-equipment supplier and user;

2 insufficient time available between ordering a machine tool and the installation date to enable a swarf-removal method to be established and implemented;

3 lack of adequate published data on swarf-handling and cutting-fluid processing equipment, stating their uses and limitations.

In this chapter an account is given of the steps that machine-tool users can take to minimise the problems of swarf removal.

COMPONENT DESIGN AND MACHINING CONDITIONS

The production of swarf is an expensive business and any possibility that may reduce the volume should be a subject for detailed examination. Normally, it is the machine-tool user who will have the task of selecting the appropriate machine and of finalising the design of the component that is to be made, and the method of processing the raw material. This means that the control of swarf is very much in the hands of the machine-tool user also and, whilst it must be assumed that the selected machine will produce the component in the best and most economical fashion, the following factors concerned with component design and machining conditions are worthy of consideration.

(a) Component design

Considerable easement of the swarf situation may be gained from efficient design of the components to be produced and, whilst every design must produce a completely functional component, close attention should be given to designing so that any necessary metal removal is held to a minimum.

Where machining cannot be avoided, attention to the following points will show benefits.

1 When castings are to be used, the minimum machining allowances must be specified on all machined faces and in bores, and the founder must be made to adhere to such allowances, commensurate with producing a good casting. Apart from reducing the volume of swarf, excessive machining allowances cost money to remove and are completely uneconomical if castings are purchased by weight.

2 Similar remarks may be applied in the case of fabrications but, here, the designer is better able to control the machining allowances when choosing plate thicknesses.

3 Bar stock is commonly used in the production of many components. The size of bar selected should always have dimensions as close as possible to the major cross-sectional dimensions of the finished component so that the minimum of metal will need to be removed and the volume of swarf produced will be restrained. Whenever it is allowable, such dimensions should remain unmachined; swarf is regularly produced in carrying out machining operations that are merely to enhance the appearance of a component without adding to its usefulness or life.

There are other considerations that should receive attention, such as the use of drop-stampings or cold-formed forgings in place of bar, the possibility of using alloy die-castings and other alternatives. All of these will assist with the swarf problem and will usually be economically desirable—see Figures 3.2 and 3.3.

(b) Machining conditions

Swarf can vary widely between very fine particles—dust in the dry condition and as sludge when wet—and long spiral strands or ribbons; in whatever form the swarf is produced different methods of handling will be required and these are dealt with in detail in the later chapters. But the machine-tool user can often simplify the swarf-removal problem by designing the tooling so that the most easily handled form of swarf is produced.

Fig 3.2 The photograph shows how swarf will collect and remain on flat machine tool surfaces

Fig 3.3 Illustrates the end of a swarf and coolant system, the elevator is collecting the large swarf chips and depositing them in a portable skip

In the first chapter of this manual the types of swarf were classified under three headings as dust or sludge, broken chips and bushy swarf. Machining methods and tooling that produce swarf as dust or broken chips may be considered as satisfactory as far as the swarf problem is concerned because both these swarf forms can be readily and efficiently handled, but the situation is quite different in the case of bushy swarf.

Chipbreaking

Bushy swarf is a nuisance, it is dangerous and difficult to handle and transport, it is not easily sold and only the lowest prices are paid for it. Every effort must, therefore, be made to avert its production. The answer will often be found in the design of the cutting tools.

It is very necessary to give close consideration to the over-all geometry of the cutting tool so that the design is such that, not only will metal be efficiently removed, but the swarf will be in the form of chips of a suitable size and shape and be propelled in the most convenient direction for removal. If this cannot be accomplished with standard tools then chipbreakers should be brought into use.

The use of cutting tools equipped with chipbreakers should be encouraged with the economics of their application taking into account the simplification of the swarf-handling problem. Chipbreakers are usually of two types; the first takes the form of a roller or some other barrier placed adjacent to the cutting edge of the tool so that the chip will impinge upon it and be forced to break away; the second and most commonly used, particularly with carbide tools, is in the form of an accurately ground groove beyond the cutting edge or, in the case of throwaway tips, a channel formed integrally. In the machining of most ductile materials chipbreakers will be successful in preventing the formation of bushy swarf and their use is strongly recommended. Some examples of correctly designed chipbreaker tools are shown in Figure 3.4.

Use of cutting fluid

The advantages, disadvantages and economics connected with the question of WET versus DRY machining are not always examined in detail. The loss of productivity in dry machining compared with wet machining is not the only criterion. With dry machining tool life may be shorter, the suppression of dust in the machining of

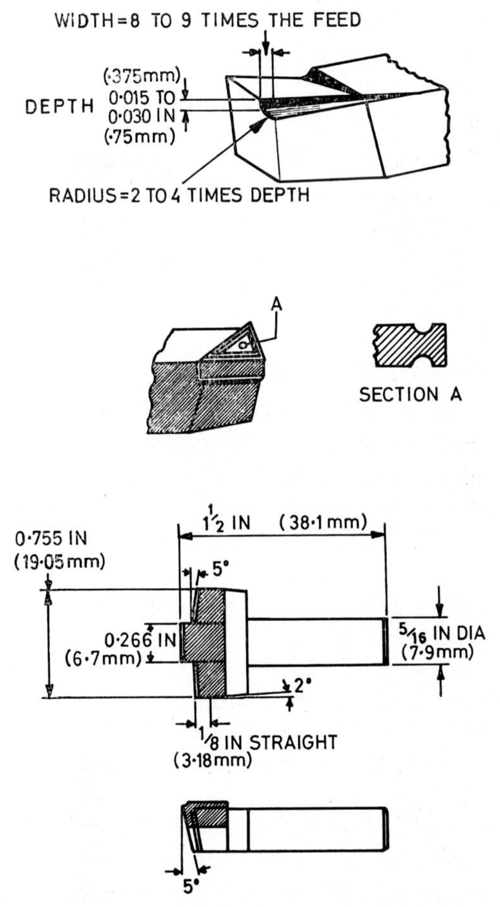

Fig 3.4 Some illustrations of typical chipbreakers

cast iron may present a problem and fixtures do not benefit from the cleaning action that the use of cutting fluid gives.

However, with wet machining the following on-costs may occur:

1 machine and local-area flooding problems that increase maintenance and housekeeping costs;

2 the cost of the cutting fluid, its filtration, effluent-disposal problems, and the costs involved in inventory, storage and handling;

3 the machine may require installing on specially prepared decks with a graded surface or kerbs, drain channels and sumps;

4 cutting-fluid life may be severely curtailed due to contamination by hydraulic and lubricating oils, necessitating early replacement;

5 where components are moving between operations requiring different cutting fluids (i.e. soluble-oil cutting fluid and honing oil) it may be necessary to introduce washing facilities to avoid cutting-fluid contamination and component corrosion.

THE HANDLING OF SWARF

By far the largest proportion of handling problems are caused as a result of long ribbons of swarf forming themselves into unmanageable bundles. These are impossible to move by hand and tend to get caught by projections from machines or by the conveying mechanism itself. The main sources of swarf of this kind are usually machining operations such as drilling, turning and boring on ductile materials such as steel, soft brass and aluminium.

If mechanical methods are to be used to remove large quantities of swarf from the cutting area of a machine tool then positive steps must be taken to ensure that the swarf is not produced in the form of continuous ribbons. It must be broken into discrete lengths—no longer than about 1 in (25·4 mm) as it comes away from the tool.

Provided swarf is broken into relatively small pieces it is not difficult to handle or convey swarf from a machine tool to a convenient take-off point. As will be seen from other sections of this manual, cutting tools can be ground to break swarf into chips and tools of this kind should *always* be used when there is a danger of producing continuous ribbon-type swarf.

The swarf-handling process can be considered in terms of three basic activities:

(a) Removal from the cutting zone to the machine periphery

Four methods of achieving this are:

1 gravity removal;
2 mechanical removal;
3 hydraulic removal by using cutting fluid;
4 pneumatic removal.

Fig 3.5 Swarf being flushed from the cutting zone by a copious flow of coolant

Generally the gravity or hydraulic approach is better as it is less costly and less prone to breakdowns but the subject is dealt with more adequately in other chapters. The critical features of swarf removal from the machine tool are the positioning of the discharge point on the machine and the steps taken, by careful design of slideways and toolholders, to ensure that swarf can freely reach this point from the cutting tool. In some cases a conveyor built

Fig 3.6 An example of the removal of swarf from an individual machine using a built-in conveyor. Note the mobile skips

into the machine may be necessary. In order to assist machine-tool suppliers in standardisation to suit all users and for the user to have flexibility to apply either above-floor or in-floor conveyors or have the machine self-contained, it is advisable to have the discharge point about 18 in (450 mm) above the floor level. For use with individual machines, a swarf container of suitable size should be positioned so that it may be handled by a fork lift or pallet truck—see Figures 3.5 and 3.6.

(b) Removal of swarf from a group of machines to a collection point

Where a group of machines in a manufacturing department is bounded by an aisle, it is usual to marshal the swarf at one or two locations to facilitate collection, mobile bins or conveyors being used for this movement. On wet cutting operations it will be necessary to separate swarf and cutting fluid at the machine or at a central location, and filter the cutting fluid for re-use. If the machines are producing components in different materials a conveying system from the machines may not be feasible as mixed swarf loses much of its value.

(c) Collection of swarf from a number of departments to a central plant location for disposal

Depending on the method and frequency of swarf sales, equipment may be needed for storage so that swarf can be sold when the price level is most attractive.

On most factory sites there is usually a space or building provided for storing swarf to await collection by a scrap-metal merchant. Depending on the swarf output of a factory the size of the facilities provided for this purpose will vary from a 3-walled brick emplace-

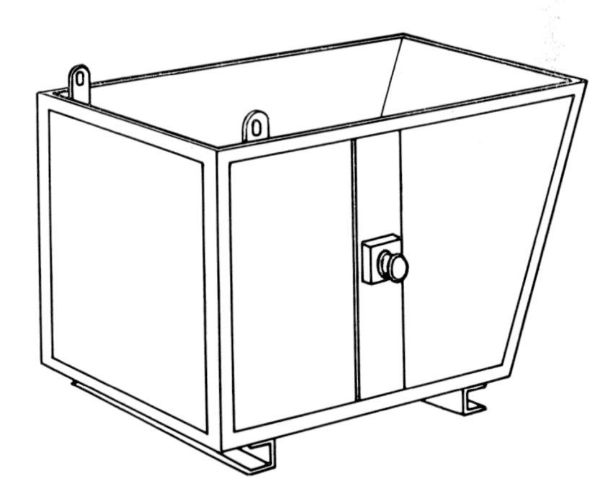

Fig 3.7 Swarf-collecting skip suitable for transporting by fork lift truck or overhead crane

ment to a well-equipped swarf house. Swarf has a significant value as scrap, however, and it is well worthwhile to consider what equipment is necessary in order to gain the maximum possible return from the scrap-metal merchant. It is also possible to derive considerable benefit from the quantities of neat cutting oil which may sometimes be retained by the swarf.

In the light of good housekeeping it is nearly always beneficial to arrange suitable methods for transporting swarf away from the machine-tool area. Removal to a central plant location, usually located at the plant periphery or outside the plant, can be handled by conveyors, fork trucks or mobile hoppers. Figures 3.7 and 3.8 illustrate typical swarf skips or hoppers. The receptacles used at the

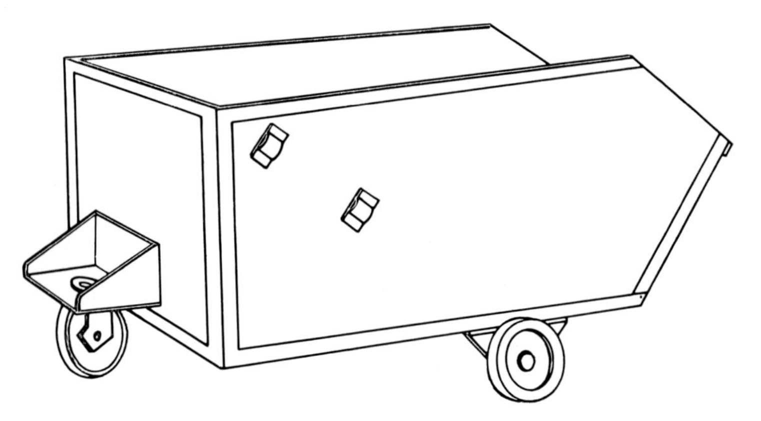

Fig 3.8 Typical mobile swarf skip that may be towed or manually manipulated

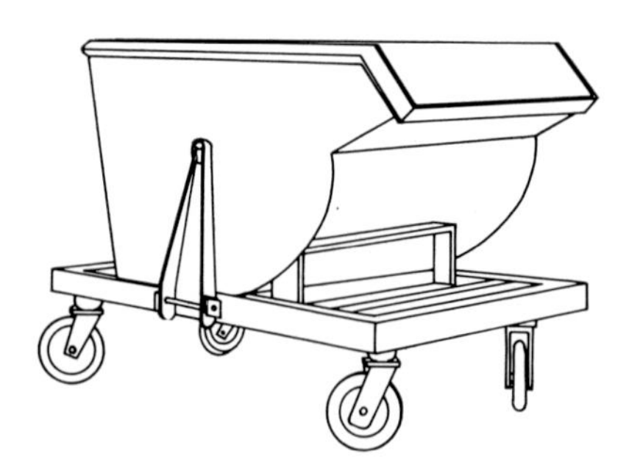

Fig 3.9 A typical tipping-type, mobile swarf skip

swarf-producing locations can, if necessary, be designed as self-tipping mobile hoppers that can be connected together and towed as a train to the swarf building.

Dry cast-iron and steel chips can be tipped directly into lorries, rail wagons, or trucks. Bushy swarf will probably be crushed into chips and wet chips passed through a centrifuge to remove cutting fluids. After processing, it is usual to collect the dry and broken swarf into an overhead hopper which can discharge by gravity into the most convenient form of transport for removal to the scrap yard or sometimes, in the case of aluminium and brass, for direct return to the foundry. Aluminium chips usually require sorting to remove any foreign material prior to bagging, and possibly crushing prior to briquetting or baling.

The need for pre-treatment before disposal is usually dependent on the nature and value of the swarf, the daily tonnage of swarf produced, a desire for good housekeeping, legal restraints and the requirements of the customers who purchase the swarf.

The cutting fluid is removed for the following reasons:

1 possible re-use, after purification, as a cutting oil;

2 for re-sale, particularly if there is a mixture of cutting fluids, to specialists who can make commercial use of tramp oils;

3 to prevent it from seeping into sewage and drainage systems as this will contravene national and local pollution acts;

4 to prevent seepage on to roadways during transportation as this could create dangerous road conditions;

5 to increase the potential selling value of the swarf.

Swarf crushing is carried out to:

1 reduce swarf to a chip size that can be handled by a centrifuge to remove cutting fluid;

2 provide a more compact load and reduce transportation costs. Crushed swarf is more easily used for subsequent re-melting operations and generally realises double the price of uncrushed swarf.

TYPES OF HANDLING SYSTEM IN MACHINING AREA

Conveying methods are classified as MECHANICAL, HYDRAULIC and PNEUMATIC; their application, limitations and design features are discussed in later chapters of this manual.

Systems can also be classified as:

(a) Single-conveyor systems

Groups of machines are linked to a single conveyor with or without a cutting-fluid supply. The conveyors may be hydraulic, mechanical or pneumatic and located above-floor or in-floor, usually limited to a 100 ft length (30 m). Where cutting fluid is used, the trough of the mechanical conveyor can act as a storage tank with a limitation of around 200 g.p.m. (9000 l min^{-1}) total flow rate. The addition of a pump, piping and a filter or strainer unit enables clean cutting fluid to be returned to the machine tools. The conveyor will be elevated at the discharge end to enable swarf to be delivered into a bin or mobile hopper.

This approach, compared with self-contained machines, reduces the number of swarf-collection points and simplifies cutting-fluid supply and replenishment problems. Compared with a central system, the equipment is relatively easy and inexpensive to rearrange in the event of plant layout changes.

(b) Self-contained machines

Swarf is collected into small mobile bins at each machine. For wet cutting operations, a storage tank for cutting fluid and, possibly, a filter is also required. With this approach each machine requires servicing by mobile tankers or barrels for removal of contaminated cutting fluid and its replenishment.

(c) Central systems

Groups of machines are linked to an integrated conveying system of mechanical, hydraulic or pneumatic type, or some combination.

The system can be used on wet (hydraulic or mechanical) or dry (mechanical or pneumatic) cutting operations.

The advantages of central systems are:

1 Central control of the cutting fluid is possible to ensure that its chemical composition is within specification.

2 Bacteriological contamination of cutting fluid can be controlled to eliminate degradation.

3 Maintenance costs are reduced since the need to service a multitude of pumps, filters and cutting-fluid tanks is eliminated.

4 The use of central systems has promoted rationalisation and reduction of a wide variety of cutting fluids that tend to be used with individual machines.

5 Generally, floor-space requirements are reduced with in-floor central systems.

6 The method of swarf and cutting-fluid collection and the reduced number of swarf-discharge locations provides for lower-cost housekeeping and improved working conditions.

The disadvantages of central systems are that:

1 They fix the positions of machine lines, thus increasing future rearrangement costs.

2 Malfunctioning of a system can result in loss of production from a whole machine line.

3 They are susceptible to being used as a disposal system for machine-shop rubbish.

FACTORS INFLUENCING SELECTION OF SWARF-HANDLING SYSTEM

Many factors influence the selection of a swarf-handling system. These include:

(*a*) plant size;

(*b*) plant layout concept;

(*c*) type of production;

(*d*) machine tool design;

(*e*) the cutting operation;

(*f*) stability of production process.

(a) Plant size

In small plants, where distances and the number of handling operations of swarf are small, the required swarf-handling method can be very different from that of a larger plant. For example, it may not be economic to use a train of mobile hoppers pulled by a tug unit, or conveyors, unless the distance between machine lines and the swarf building exceeds, say, 250 ft (80 m). For shorter runs, a fork truck may be more economical and, in a very small plant, the use of manually pushed hoppers may be the solution.

(b) Plant layout concept

Plant layout of machine areas can be based on the concept of (1) PRODUCT or (2) PROCESS.

The BY-PRODUCT layout groups unlike machines in one area to produce a product or similar product range. From the swarf handling viewpoint it has the advantage of common swarf material. However, if the machines range from, say, roughing lathes through to finish grinders or honers, the nature of the swarf will be variable (from spiral coils to sludge) and various cutting fluids may be required. BY-PRODUCT layouts are usually associated with mass production, and therefore high swarf generation, which may justify the application of sophisticated central systems for swarf and cutting-fluid handling. However, a number of small systems, as opposed to one large system, may have to be employed. For example, it is not normal to feed sludge from a group of grinding machines into a system covering, say, roughing operations. A separate system or method of collection should be installed.

The BY-PROCESS layout has the advantage of grouping like

machines together and thus probably generating similar swarf, but this is advantageous only if ALL the component materials are similar. If materials as dissimilar as ferrous and non-ferrous are cut, this will prevent the use of conveyors or central systems. Also, the cutting fluids may differ if component materials differ. Conveyors could be applied to a BY-PROCESS layout to handle a variety of component materials, providing batch runs are long enough.

(c) Type of production

Types of production are usually classified as (1) UNIT, (2) BATCH, (3) FLOW. The effect of this on swarf handling is closely linked with the plant-layout concept as, usually, UNIT and BATCH type production is carried out on BY-PROCESS layouts and FLOW production on BY-PRODUCT layouts.

In unit production a machine can be used for producing a wide variety of components from different materials and requiring a variety of cutting fluids. This necessitates the use of a flexible swarf and cutting-fluid-handling method which would be individually designed to suit the machine tool. Similar comments would apply to machine tools used for batch production unless the batch runs were consistently long. The constraints imposed by flow production have been dealt with in the previous section.

(d) Machine-tool design

The position of swarf and cutting-fluid outlets is the critical factor here, together with whether the swarf will discharge continuously or require regular manual assistance. Where batch swarf removal is necessary, the mechanical conveyor for collecting from a number of machines is more reliable as bulk dumping jeopardises the operation of a hydraulic system.

Machine-tool design modifications can exert little influence on existing machines, but close liaison between the USER, MACHINE DESIGNER, and EQUIPMENT DESIGNER can be advantageous on new machines. Therefore, the timing of finalising the swarf-handling method is vital and should be fitted in between the machine-tool procurement date and finalisation of preliminary machine-tool design.

(e) The cutting operation

The factors in the cutting operation that affect the selection of a swarf-handling method are:

1 type of component material;
2 rate of chip generation;
3 nature and structure of the swarf (as indicated in Chapter 1);
4 type of cutting fluid and flow rate.

(f) Stability of production process

Past experience and forward planning will indicate the level of stability of the production process. Model changes, component-design changes and sales-volume changes often necessitate layout changes. The installation of a non-flexible swarf-removal method demands long-term stability of layout to ensure economic returns on investment and to avoid costly rearrangement of the equipment installed.

THE METHOD-STUDY APPROACH

The final section of this chapter is intended as a tool to enable a critical examination to be carried out in a logical and methodical sequence.

The earlier sections of this chapter and the following chapters should provide sufficient information to result in a method that is appropriate to the operational situation and environment of the plant involved. Dependent on the circumstances, it may be necessary to consult frequently with equipment suppliers during the study period. To ensure that the final selection of a swarf-handling system is arrived at in a satisfactory manner it is recommended that the method-study sequence of

SELECT—RECORD—EXAMINE—DEVELOP—INSTALL—MAINTAIN

be employed to enable a critical examination of all information to be undertaken in a logical and methodical manner.

The swarf problem under review can vary from:

1 dealing with one manufacturing department to dealing with all departments in a plant;

2 establishing a new swarf-handling method for an existing plant to establishing a first swarf-handling method for a new plant.

In a new plant, the planning engineer has the advantage of being able to influence the plant layout and is not involved in the disposal of assets connected with an existing swarf-handling method, but in an existing plant the constraints of layout and available assets may necessitate compromise solutions.

It may be found that some of the methods of analysis and questions are not applicable to the peculiarities of the study in hand and it may be necessary to substitute questions to suit the particular problem. Furthermore, during most projects it will be necessary to enlist the support of equipment and machine-tool suppliers at various study phases.

(a) SELECT

Area of study For a new plant, selection is not involved and the task of establishing new facilities is mandatory. Where an existing plant is involved the planning engineer should look for substandard conditions and assess whether a change will achieve sufficient results to cover the cost of a study. To ensure subsequent co-operation in changed methods the ideal first problem to be tackled is the one that can be introduced quickly and economically and requires minimum disruption of established working methods.

Scope of the study The scope of the study is to establish a number of alternative plans for (1) a new plant or (2) improving methods in an existing plant and selecting the best method of implementation.

Define objectives It is important to establish the objectives of a swarf study in terms of reasons for the action and the results expected after implementation. It is against these objectives that alternatives will be compared and the subsequent installation will be measured and controlled.

(b) RECORD

This is the vital stage of obtaining and recording information that is relevant to the study; on the accuracy of this information will depend the validity of the generated alternatives and the subsequent selection and installation. The basic tools for recording are (1) copies of the detailed plant layout, (2) pro-forma charts for the collection and recording of data.

Separate charts should be prepared for each manufacturing department. If there is a variety of operations in each department (such as a turning section and a grinding section) separate charts should be prepared for each section.

Standard information on the chart would be:

1 component description;

2 component material;

3 weight of swarf (i.e. weight of component on input—weight on output);

4 hourly production rate;

5 hourly amount of swarf = (3) × (4);

The second part of the form would contain information against each machine:

6 asset record number of machine;

7 name of machine;

8 type of swarf (bushy—broken—dust or sludge);

9 type of cutting fluid;

10 amount of cutting fluid (g.p.m.);

11 standard of cutting-fluid cleanliness required;

12 position of swarf and cutting-fluid outlet;

13 operation number of machine;

14 type of cutting operation (i.e. grinding, tapping, honing, etc.)

The layout and charts should be cross-referenced so that, during layout review, the factors associated with each machine can be related to the layout position of the machine.

(c) EXAMINE THE FACTS CRITICALLY

This stage concerns the analysis of the facts obtained and the generation of alternatives, together with a record of the implications of the alternatives. Attitude of mind is important at this stage and it is necessary to examine the facts as they exist and not as they are thought to exist. Also, preconceived ideas should not be allowed to influence analysis; the facts must be analysed in a logical, challenging and critical manner. The general questions listed in the following chart will require framing to suit the specific study involved and will also assist in looking for relevant information required for the analysis—see Table 3.1 below.

During the analysis of current methods and the generation of alternatives, the following approach should be taken:

1 Challenge the purpose of any activity or feature to see whether it can be *eliminated*. For example: why should the cutting operation be wet—can cutting fluid be eliminated?

2 Where an activity cannot be eliminated can it be *changed or combined*?

3 Can any activity be simplified?

4 Are there any future developments (production increases, process changes, material changes) that may affect the study?

The generation of alternatives is also assisted by examining the types of equipment that will suit the peculiarities of the swarf problem under review; information on this is supplied in the following chapters. Table 3.2 provides a guide on establishing alternatives, the positions of X indicating equipment approaches that are *not* feasible for application to certain swarf problems.

TABLE 3.1 *Examples of the kind of questions, listed in order of importance, that must be answered when considering new swarf-handling equipment**

	Information on current method	*Analysis of current method*	*Generation of alternatives*	*Selected alternatives*
Purpose	What is done?	Why is it done? Is it a must or a want?	What else could be done?	What should be done?
Place	Where is it done?	Why there? Advantages Disadvantages	Where else could it be done? Advantages Disadvantages	Where should it be done?
Sequence	When is it done? After Before	Why then? Advantages Disadvantages	When else could it be done? Advantages Disadvantages	When should it be done?
Person	Who does it?	Why that person? Advantages Disadvantages	Who else could do it? Advantages Disadvantages	Who should do it?
Means	How is it done?	Why that way? Advantages Disadvantages	How else could it be? Advantages Disadvantages	How should it be done?

* This chart is established on the premise that an existing method is in operation. For a new plant it would be necessary to start at the section of generation of alternatives. One method would be established and further alternatives would be generated by using the questioning approach stated. A wide range of alternatives from fully manual to fully automatic could be generated but would probably prove excessive. The size of the swarf problem and the plant are the only guides as to where to start.

TABLE 3.2 *Method of choosing a swarf-handling system by a process of elimination*

Component material / *Equipment*	*Cast iron*		*Steel and non-ferrous*			
	Wet	*Dry*	*Bushy*	*Broken*	*Sludge*	*Dust*
1 Pneumatic	X		X	X	X	
2 Hydraulic		X	X			X
3 Mechanical						

In the majority of factories in which large quantities of swarf are produced it will nearly always be necessary to consider handling swarf by one of the following methods:

1 *Pneumatic*—A collection system can be connected to a large number of machines as a central system, or a number of small collectors can be fitted to individual or pairs of machines.

2 *Mechanical and hydraulic*—Can be connected to a small group of machines as a single-conveyor system or to a large group of machines as a central system consisting of 2 or more interconnected conveyors or a combination of both types.

The alternatives of using mobile bins or conveyors to connect all manufacturing departments will require examination. Mobile bins are less costly and more flexible and can be moved in trains by a tug unit (for a large plant), by fork truck, or manually in the smaller plant.

(d) DEVELOP

This stage requires the preparation of capital costs, operating costs and the implications of the various alternatives to enable them to be compared and subsequently a selection (i.e. the plan of action) to be established.

Capital costs For each alternative the procurement and installation cost is required for comparison.

Operating costs The operating cost for each alternative is required in great detail. This is to ensure that a scheme is not solely selected in favour of low capital costs as, generally, such a scheme will involve higher operating costs. Factors such as direct labour, indirect labour, maintenance, floor space, general services, etc., provide a realistic assessment of total operating costs.

Economic selection A company's accountant is normally required to carry out an investment appraisal to determine which scheme offers the greatest economic advantage to the company.

Final influencing factors All other factors should be reviewed as these may have the effect of requiring a less-economic scheme to be selected.

(e) INSTALL

This stage covers:

1 Management approval of the scheme and the allocation of funds.

2 Implementation of the scheme involving:

(*a*) procurement of equipment;
(*b*) preparing for the change;
(*c*) installation of equipment.

Management approval The presentation of the case to management is a vital point. Management should not be asked only to concur with the recommendation in isolation but should be requested to make a decision which involves agreeing that the recommendation is the best choice from the alternatives reviewed. The presentation should be prepared in the same sequence as the method-study approach.

Procurement of equipment The issuing of enquiries to equipment suppliers to obtain competitive tenders must be preceded by the vital stage of specification preparation. To enable suppliers to quote fully to meet the requirements and ensure comparative tenders, the specification should provide information on:

1 What is the swarf problem?—provide plant layouts and the information charts mentioned in stage 2: record.

2 What the requirements are to solve it?—the recommended method.

3 What constraints must be imposed on the supplier (site conditions—timing—standards, etc.)?

A detailed specification is essential but should not prohibit individual suppliers from recommending alternative ideas. This can be arranged in two ways:

1 Request that all suppliers quote to specification, to provide equal comparison for selection, and then quote separately and additionally their alternative ideas.

2 Let the suppliers quote differently to accommodate their alternative ideas.

The first method creates more tendering work for the supplier but usually ensures a just selection of the successful bidder.

Preparing for the change All personnel in the plant who are directly, or even indirectly, involved in the proposed swarf-handling method should be consulted. Their full understanding and co-operation is essential to ensure that the swarf scheme will operate successfully. For staff directly involved in the more complex parts of the scheme some form of training programme may be essential.

Maintenance staffs should be encouraged to meet the supplier and subsequently, perhaps, to work with the supplier on installation to ensure a complete understanding of the equipment. Visits to the supplier's works at various stages of construction will also assist maintenance staff to ensure subsequent continuity of equipment operation. During the manufacturing and installation stage the supplier should be requested to supply an operation and maintenance instruction manual. This should include such things as starting-up and shutting-down procedures—preventive maintenance schedule—drawings—parts lists—recommended spares holdings, etc.

Installation The planning engineer will be involved in checking the installation from a timing and engineering-standards viewpoint, and an operating validation period will be required before final acceptance of the equipment.

(f) MAINTAIN

The new swarf-handling scheme must be controlled when in

operation. The planning engineer should check the operation at regular intervals by measuring actual performance against the budgeted factors of the study.

Problems may require resolving that will create changes in the scheme. Also, improvements to the scheme may become apparent and it may be possible to incorporate them; if this is not possible, the information should be recorded for use with future schemes.

4 MECHANICAL METHODS FOR THE CONVEYING OF SWARF

INTRODUCTION

Mechanical conveyors have three advantages that favour their use in the transporting of swarf. These are:

1 They are generally robust in design and the materials used in their construction can be tough and hard wearing.

2 The operation of mechanical conveyors is unaffected by the presence of cutting fluid and they can be fully integrated with hydraulic conveying systems.

3 They can be used effectively in horizontal and inclined planes and, in some cases, in positions approaching the vertical.

There are many different types of mechanical conveyor and several typical examples are illustrated and their main characteristics are given later in this chapter together with a description of their main uses in dealing with the problem of swarf transportation.

MECHANICAL CONVEYING AND TYPES AND SIZES OF SWARF

One of the major difficulties to be overcome in any system for the handling of swarf in bulk is the wide variation in the size of the swarf, from the fine dust from brittle materials to the long helical ribbons formed during the turning of steel. A very good indication of just how bulky swarf can be is given in Chapter 2 where the volume of swarf produced in machining one cubic inch of various materials is illustrated, together with some of the types of swarf that may be encountered. It is because of these wide variations that not all conveyors are suitable for transporting all types of swarf and that selection of the most appropriate equipment calls for close consideration. Each of the three main types of swarf has its own peculiarities that affect its transportation properties and it is worthwhile considering these in some detail.

Dust in the form of granular particles may be produced during the dry machining of most brittle materials, particularly cast iron, and will be intermingled with the normal swarf chips that flow from the cutting tool. This dry dust will tend to accumulate and to be difficult to move even though the chip content of the swarf is being successfully handled; moreover, because of its fineness, it is quite likely to enter confined spaces and clog the mechanism of the conveying equipment or lead to excessive wear. Grinding and honing operations also involve the production of dust but in this case it will almost certainly be contained in a cutting fluid and will form a sludge or slurry. Swarf in this form may contain other particles of contaminant such as grinding-wheel grit, wheel-bonding material, etc., all tending to form a mixture that does not present a straightforward conveying problem. All these factors must be recognised and assimilated when conveyor types are being considered.

The middle class of swarf, which in this manual is designated as 'broken chips', may include sections of broken ribbon helices often one full turn in length as would come from a turning operation, chips produced during planing, shaping or slotting operations, or splinters as might be formed during a milling operation. It could well be that the swarf created on a horizontal boring machine would contain a mixture of all the three swarf shapes just mentioned, thus adding to the difficulties attendant on removing and transporting. To add complications to the methods of dealing with this class of swarf there is the fact that it may be the result of either wet or dry machining. In the dry state it may flow quite freely or, if broken helices are present, then it may become mechanically interlocked and build up into a mass, thus adding to the problem of conveying it. If, however, the cutting has been done with the aid of a cutting fluid then a fresh set of conditions is presented. Straight cutting oil has a moderately high viscosity and splinters or small chips having absorbed such oil will tend to become glued together and form a mass that is not easily moved, or a coarse slurry of high viscosity having properties of strong adherence. Conveyors that must operate under wet cutting conditions will have to deal not only with swarf but with the effects of cutting fluids also and such conditions warrant very careful investigation.

The third and most difficult class of swarf is that known as 'bushy', the type of swarf resulting mainly from the rotary machining of ductile materials using cutting tools that are not provided with chipbreakers. Under either wet or dry machining conditions, the long ribbons or helices that are continually produced whilst steel is being turned will rapidly form into a tangled mass that is not easily cleared from the cutting zone and the machine and, therefore, must be a problem to transport. Even when manual methods are used in dealing with swarf of this nature a number of difficulties is experienced in handling and manipulating it and it follows that mechanical transporting will not be a simple matter. Bushy swarf is also dangerous; the edges are usually quite jagged and extremely sharp and can quickly cause a nasty cut unless great care is taken. If the swarf is coated with cutting fluid that is contaminated then cuts and wounds may quickly become septic. Such swarf as this will rapidly extend beyond the confines of a machine tool and become a hazard upon the workshop floor; it may wrap around feet and legs, cutting footwear and clothing, and it will deposit cutting fluid on the floor, making for slippery conditions and adding to the possibility of drains becoming fouled. From the foregoing, it will be realised that bushy swarf will need very special consideration if it is to be successfully handled and transported.

MECHANICAL CONVEYORS FOR TYPES OF SWARF

A number of the problems involved in the transportation of various types of swarf having been outlined in the preceding sections, this section will give a general indication of the type of mechanical conveyor best suited to a particular type of swarf. Without going into full details, the following may be used as a guide:

1 Swarf in the form of dust or sludge may be successfully handled by drag, scraper or flight conveyors but, because of the penetrative and abrasive qualities of this swarf, the fewer the number of working parts in contact with the swarf the longer will be the life of the equipment. Magnetic conveyors will deal very satisfactorily with ferrous swarf but will not extract other contaminants from the cutting fluids. If the swarf is in the form of dry dust, then bucket elevators and vibratory and screw conveyors will all be found to be efficient.

2 Swarf in the form of broken chips may be successfully transported by almost any type of conveyor with the exception of those that reciprocate and use forks as distinct from paddles to push the swarf along. Before equipment is ordered, the advice of conveyor manufacturers should be sought on any proposed application. Belt conveyors have and can be used for specific applications but care must be taken to ensure that the feed on to the conveyor does not permit swarf to spill over the edge of the belt.

3 For the conveying of bushy swarf in the horizontal plane, the reciprocating type of conveyor is very effective when equipped with either forks or paddles. A form of this conveyor is now on the market that allows swarf to be transported around bends having included angles of up to 90°. For horizontal movement and at reasonable inclines, bushy swarf can also be carried on conveyors on which connected slats form a continuous steel belt. Very often brackets or prongs are provided at intervals and at right angles to the direction of travel to prevent the swarf from slipping backwards on the inclines.

INCORPORATING THE CONVEYOR WITH THE MACHINE TOOL

In order that swarf may be dealt with at the nearest point to its creation it is sensible to build the conveying equipment into the machine tool, particularly on transfer lines and multi-station machines where considerable quantities of swarf are produced. And, as more engineers become aware of the magnitude of the swarf problem, it is likely that there will be an increasing demand for machine tools with integral swarf-removal arrangements—see example in Figure 4.1.

Fig 4.1 An illustration of a screw conveyor extracting swarf from a bar automatic. Such conveyors will transport the swarf at speeds of up to 10 feet (300 cm) per minute

The first problem to be overcome will be that of the size of the equipment that is to be incorporated. Whatever type of conveyor is selected, space will need to be made available within the structure of the machine and in such a way that the rigidity of the machine remains unaffected. Where individual machines are concerned, this could mean a completely new concept in machine-base design and a complete breakaway from the present conventions. On multi-station or rotary transfer machines the problem is twofold; not only must the necessary space be made available but arrangements must be made so that the swarf produced at each station is directed to the extraction unit. Because of the design of the main bed of a transfer line, the problem may not be so acute but, without detracting from the stiffness of the bed or interfering with the operation of the work-pallet transfer mechanism, space must still be found.

Another important factor for the designer to take into account is the amount of swarf that it is anticipated will be produced, as this will again bring in the question of conveyor sizes. A transfer line may be anything from fifty to sixty ft long and have twenty or more machining stations, each producing swarf that needs to be continuously cleared. A situation such as this will very often call for a conveyor running under the full length of the transfer line, and it becomes necessary to make a compromise between the over-all width of a conveyor and the maximum width of opening which can be reasonably included in the structural design of each unit of the transfer machine.

Mechanical conveyors are not immune from mechanical troubles and need to be regularly serviced in view of the difficult conditions in which they operate. The machine-tool designer must bear this in mind when planning the location of the conveyor within the machine structure, especially so in the case of a long transfer line where the grouping of the machining stations along both sides of the main bed will make access to any part of a conveyor a difficult operation. The problem of providing reasonable access to a conveyor for the purpose of maintenance is, perhaps, even more acute in the case of rotary transfer machines. Because of the compactness of this type of machine it is not always easy to get into the base without dismantling a large proportion of its main structure. It is often beneficial, therefore, to consider the possibility of making use of flange-mounted, screw- or slat-type conveying units which are not only capable of withdrawing swarf but which can also be positioned in such a way that the entire unit may be removed easily for maintenance purposes.

Machine tools that warrant the inclusion of a swarf-removal unit will probably operate at high production levels, producing large quantities of swarf and using high volumes of cutting fluids for cooling purposes and to flush away the swarf from the cutting zone. This presents another problem to the machine-tool designer, who must now select the conveyor that will operate satisfactorily in the presence of quantities of liquid without allowing the liquid to leak or otherwise find its way on to the workshop floor, and he must ensure that the materials from which the conveyor is constructed cannot be affected by contact with the cutting fluids.

The foregoing paragraphs illustrate just a few of the pitfalls that the designer must avoid if his machine tool is to give satisfaction to the ultimate user, and close and complete liaison with the conveyor manufacturer is stressed at all stages of the development.

SWARF CONVEYORS AS ACCESSORIES TO EXISTING MACHINE TOOLS

Having given some consideration to the problems associated with an integrated swarf-removal system, reference must now be made to the factors to be taken into account when it is desired to mechanise the removal of the swarf from existing machine tools.

The first problem to be overcome is the fact that the structure of the machine tool does exist and cannot be changed to allow for the entry of any form of conveyor. This means that a design of conveyor must be selected that will fit into the existing structure or which can be threaded into position through apertures in the base of the machine at appropriate points.

The second problem is to ensure that the portion of the conveyor that will protrude from the machine will allow the necessary clearance to working parts of the machine, for instance, the saddle of a lathe. This leads to a secondary consideration; as the machine-tool layout is in existence, the choice of a conveyor may be influenced by the amount of space available behind and around the machines.

And thirdly, the width of the conveyor must be such that it will collect the swarf as it falls from the cutting zone. This can present a difficulty with some types of conveyor when the cutting tool traverses laterally over an appreciable distance and the swarf falls away from the cutting zone throughout this traversed length.

Having considered these and other associated problems, it should be realised that they can all be overcome and the ultimate benefits to be gained will outweigh the initial difficulties. There is a number of mechanical conveyors on the market today which can be readily applied to individual machine tools. The design is usually such that, when the swarf has been collected as it falls from the cutting zone, it is elevated for discharge into a suitably located skip or on to a cross conveyor serving a number of machines. Depending on the type of swarf, these conveyors may be of the magnetic, drag-link or continuous-slat type, all of which are illustrated in this chapter.

THE ECONOMICS OF MECHANICAL SWARF CONVEYING

The removal of swarf from machine tools in the great majority of small workshops is still performed by a labourer equipped with a rake, shovel, brush and barrow. In many ways this has been a reasonably cheap and efficient method but the availability of such unskilled labour is lessening and labouring costs are constantly rising. Manual removal also has a number of drawbacks apart from the time consumed, some of which may be summarised as follows:

1 When the labourer is occupied with other work, swarf may gather in quantity on some machines, making it necessary for the operator to stop production whilst he clears the swarf to the floor or whilst he stands by awaiting the arrival of the labourer.

2 The labourer may find it difficult to clear away the finer particles of swarf and these will be re-circulated in the cutting fluid to the detriment of the machine, workpiece surface finish and possibly the life of the cutting tool.

3 However carefully the labourer deals with the swarf it is very likely that some of it will find its way on to the workshop floor where it will constitute a hazard to the well-being of the personnel. It is also almost inevitable that some of the cutting fluid will be deposited on the floor, and, as has been said before, this can render floors slippery and dangerous and some of the fluid can find its way into the drains, causing fouling.

4 Often, with manual removal, the swarf must be taken by barrow for dumping in a designated spot on the ground in the works yard. When sufficient swarf has been collected it will need to be handled

again when the scrap-metal merchant is prepared to take it away. When the swarf contains cutting fluid the practice of dumping it on the ground must be stopped in view of recent legislation governing the disposal of industrial effluents and the more stringent attitude being adopted by local authorities.

Forward-looking managements will be well aware of all these potential and actual costs and will be considering the economic factors involved in a change to mechanisation.

The first consideration must be the capital cost of the necessary equipment together with any installation costs. Generally speaking, mechanical conveying equipment is quite reasonably priced, even if an elaborate installation is required. Mechanical conveyors can generally be installed above floor level, thus saving expensive excavation of the floor and leaving the machine-tool layout quite flexible.

Because of their robust construction mechanical conveyors have a long and useful life, and because they are so simple they will cause little trouble and require few replacements, given reasonable maintenance attention. They are also relatively economical to run compared with other types.

For the removal and handling of swarf, most mechanical conveyors will be required to operate at medium-to-low speeds only, 10 ft (300 cm) per minute being quite a normal speed for individual units. Even with complex systems serving numerous individual machine tools or several transfer lines, taking into consideration the weight of the swarf and the low conveying speed, the horse-power consumption will be very low.

Swarf that is completely collected and transported by mechanical conveyors need not be manually handled, thus saving labour costs. With a cover-all system, the swarf can be carried and deposited into elevated hoppers which can be arranged to discharge directly into lorries or railway trucks. If the swarf is to be passed through a crusher, it can be fed from the conveyor into the crusher and then elevated to hoppers or, if the cutting fluid is to be extracted, the conveyor can deposit it into the separating equipment.

These are just a few of the benefits and economies that will follow from the installation of mechanical equipment for removing and transporting swarf, and there is no doubt that other advantages will be realised once a system becomes established.

DETAILS OF PARTICULAR CONVEYORS

Drag, scraper or flight conveyors

These types of conveyor are well suited to deal with swarf in the form of broken chips or dust, whether the swarf be wet or dry. The scraper type of conveyor is usually used for applications where the width is over 10 in (254 mm) and the drag-link type for widths below this dimension—see Figure 4.2.

The scraper conveyor usually incorporates two malleable-iron chains with scrapers or flights mounted at regular intervals, whilst the drag-link conveyor usually has only a single strand of malleable-iron chain with each link having a wing which drags the swarf along. In both cases the chains, when conveying, slide along renewable wearing plates and on the return side they may be

Fig 4.2 An illustration of the links in a drag-type conveyor, specifically designed for swarf removal

carried in channel sections or slide on the top of the conveying side of the chain. The cast links may be joined together by link pins or may be made with hooks at one end which slip over bars at the other end, rendering the chain more or less self-cleaning and reducing the risk of build-up and locking.

These conveyors may be inclined at the delivery end to allow the swarf to be deposited into a skip or hopper—see Figure 4.3.

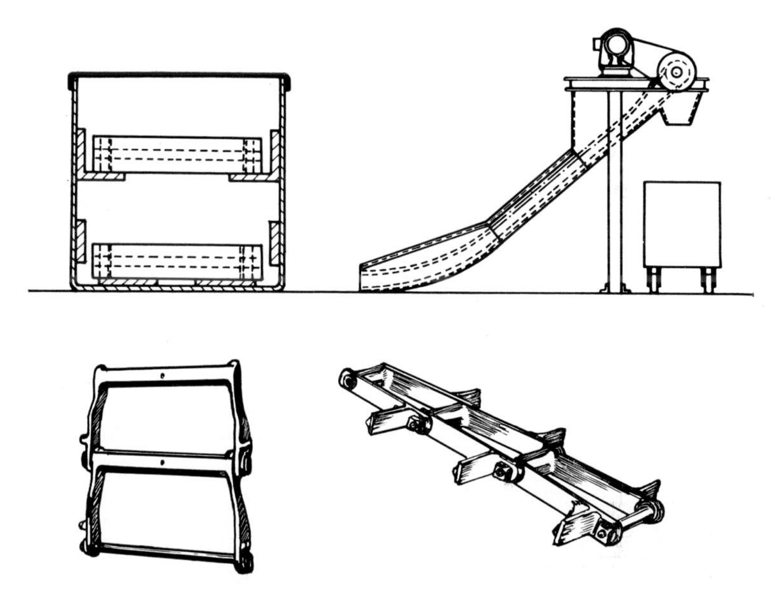

Fig 4.3 A drag-link scraper conveyor. Two types of links are shown, one of them incorporating projecting 'flights'

Pan conveyors

As the name implies, this type of conveyor consists essentially of a number of pressed-steel pans which are attached to two strands of roller chain so that they overlap. The chain rollers are usually fitted with bushes and are supported on cross-spindles whilst the pans have a number of small holes provided in the bottom so that the cutting fluid may drain from the swarf into the containing trough and find its way back to the sump. The trough is fitted with continuous valances along each side and these act as screens to protect the rollers and bearings from falling swarf. Conveyors of this type are suitable for mounting at floor level or for letting into the floor; the head can be inclined at angles up to 20°, allowing for discharge into a hopper, and broken swarf chips and small amounts of bushy swarf are efficiently transported—see Figure 4.4.

Piano-hinge conveyors

This is a good general-purpose conveyor that is well suited to the conveyance of broken swarf chips and will, in some cases, deal

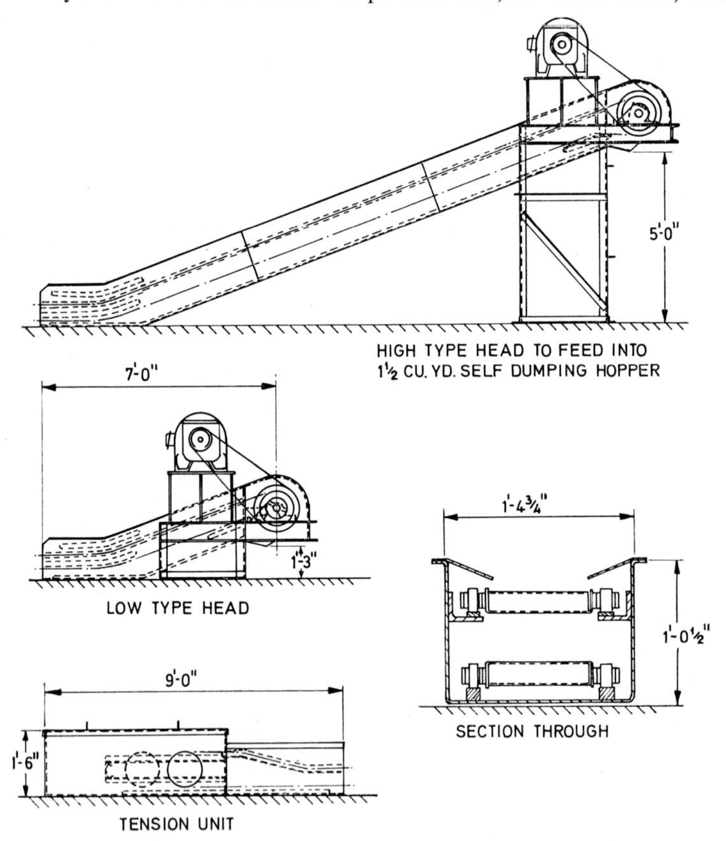

Fig 4.4 This shows the design of a typical pan-type conveyor. The two delivery heads are identical but they can be positioned at any reasonable height above floor level

adequately with bushy swarf. The tray sections are again carried on two bushed roller chains and the roller spindle extends through the width of the trays to form the hinge pin, the edges of the trays being folded to make the hinges. Each tray section is provided with side valances and these overlap when the sections are assembled, thus presenting a continuous moving trough. When wet swarf is to be conveyed, holes are provided in the bottom of the tray sections to facilitate the drainage of the cutting fluid. Generally, this type of conveyor can be used at a more acute angle than the pan type and is obtainable in greater widths—see Figure 4.5.

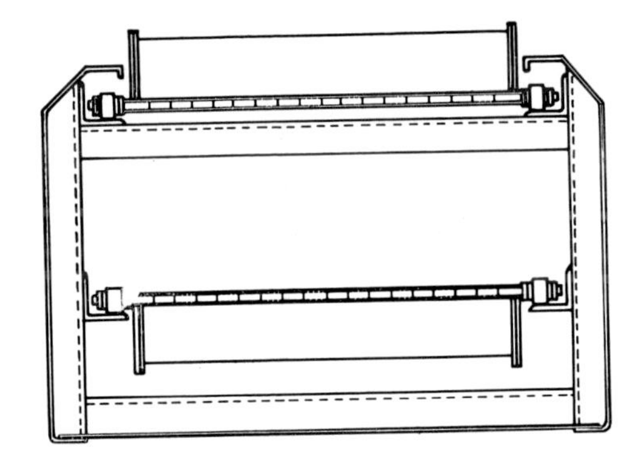

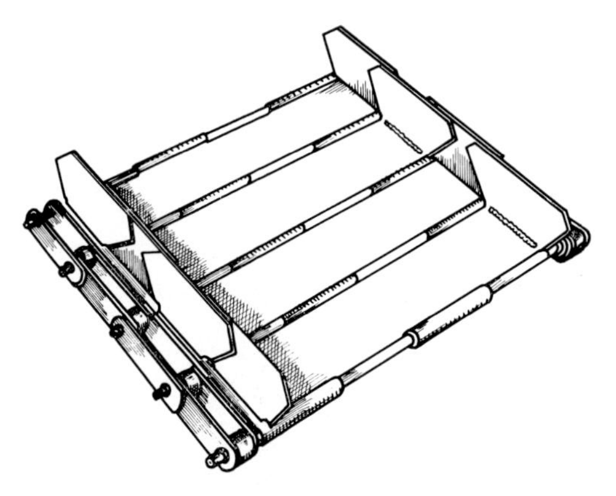

Fig 4.5 This illustration is of a piano-hinge-type conveyor. The upper view is sectional through the body of the conveyor and the lower view clearly shows the assembly of the carrying segments

Magnetic conveyor and elevator

This is a relatively simple piece of conveying equipment that may be used in the horizontal plane and at quite steep angles to the horizontal, and it is normally used for the removal of large swarf chips from the cutting fluid. There are two types available; in the first, permanent-magnet assemblies move under a non-magnetic skin and in the second type, a belt of suitable material moves over a stationary magnet system. In both cases, the magnets attract the magnetisable swarf particles, enabling them to be extracted from the cutting fluid and carried away to the discharge point. This type of conveyor may be conveniently incorporated into a machine tool; it can deal with up to 60 g.p.m. (270 l min^{-1}) of cutting fluid and will extract up to 5 lb (2·30 kg) per minute of swarf—see Figure 4.6.

The main disadvantage of this type of equipment when used for extracting the swarf from the cutting fluid is that only the ferrous

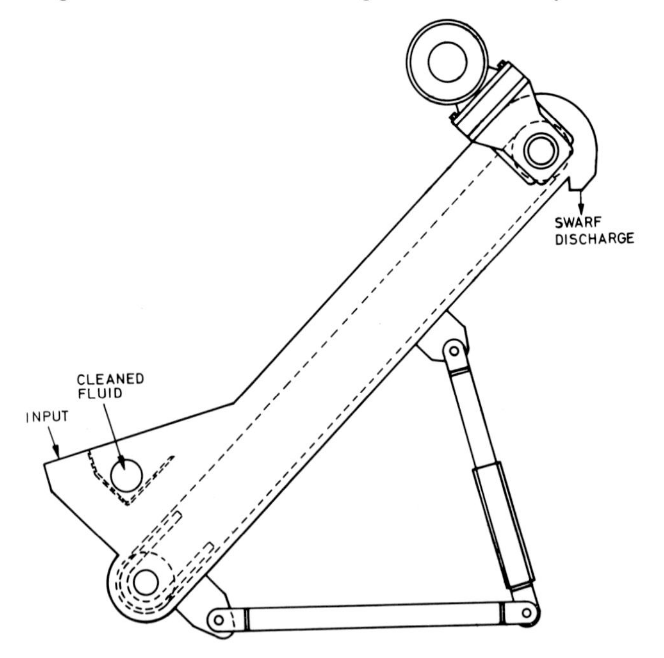

Fig 4.6 A combined magnetic conveyor and elevator

content will be taken out. Units of this type are often used on grinding machines and they deal effectively with the swarf sludge, but there will always be a proportion of non-ferrous contaminant remaining in the cutting fluid.

Overlapping tray conveyor (sealed-slat conveyor)

This is a most useful type of conveyor and operates with equal efficiency with either broken chips or the bushy type of swarf. The carrier consists of a number of slats which are mounted on two lengths of bushed roller chain, the rollers having flanges for centralising purposes and the whole forming a continuous steel belt. The trays are overlapped at the pitch points of the supporting

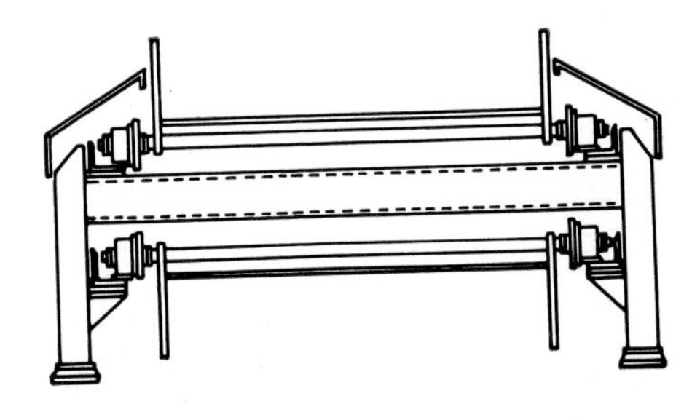

Fig 4.7 This is an overlapping-tray conveyor. The lower view shows the assembly of the trays and the rollers

chains so that there are no gaps along the carrying surface. Overlapping side valances on the trays prevent spillage and assist in creating a continuous trough. The head of the conveyor may be inclined for the purpose of discharging, and lengths of steel angle section can be fitted to the trays at intervals to assist in carrying the swarf when the angle of inclination is appreciable.

This type of conveyor is usually obtainable in widths between 18 in and 60 in (450 and 1,524 mm)—see Figure 4.7.

Swarf extractor/conveyor

This is a variation of the piano-hinge type of conveyor, adapted for use as a single unit with individual machine tools. The hinged steel slats that make up the simulated continuous steel belt have interlocking side wings, and lengths of steel angle section may be attached to the slats at intervals to assist the transporting of the swarf on inclined sections. By maintaining the width of the slats at the minimum, say 2·5 in (63·5 mm) approximately, it is possible for the full depth of the conveyor casing to be accommodated in the

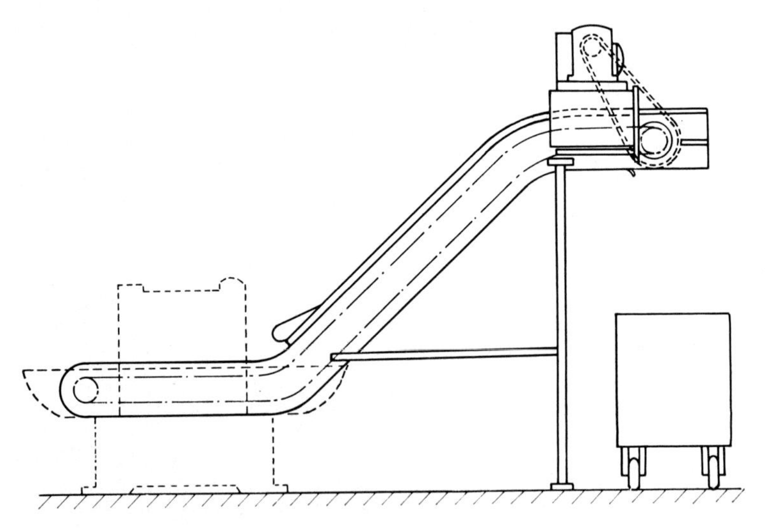

Fig 4.8 A combined swarf extractor and conveyor of a type that is suitable for use with certain individual machine tools

cutting-fluid tray of many types of machine tool. Although this conveyor is most effective when the swarf is in chip form, a reasonable proportion of bushy swarf can be carried and, if a top cover is fitted, this bushy swarf will be compressed by passing under it. Drain holes can be provided in the slats and the casing to allow the cutting fluids to return to the sump—see Figure 4.8.

Screw conveyors

This type of conveyor, also known as a worm conveyor, is suitable for the transportation of most kinds of swarf but is most effective when dealing with swarf in chip form or as dust, in either wet or dry condition.

Essentially, the conveyor consists of a trough or casing containing a longitudinal central shaft fitted with spiral blades that convey the swarf along the trough as the shaft rotates. The blades may be in the form of a continuous-strip helix, with constant or varying pitch, secured by welding to the central shaft, or separate blades of paddle or crescent shape attached to the shaft by bolting. Screw

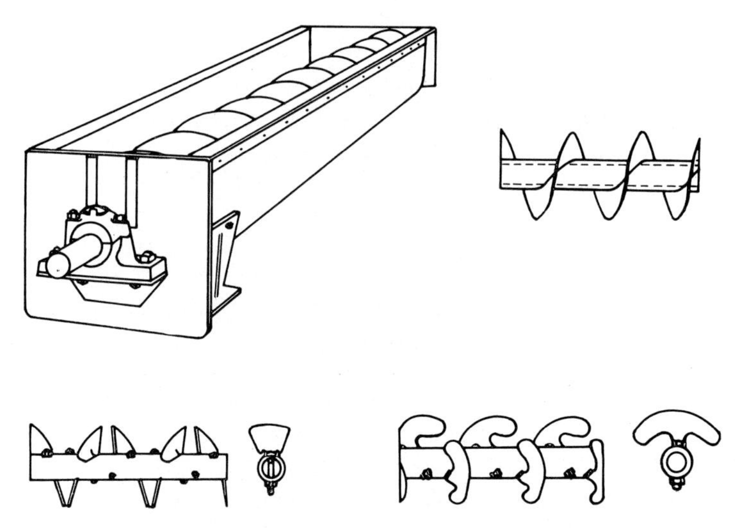

Fig 4.9 A typical worm or screw conveyor contained in a trough. Three types of blading are shown

conveyors are made with diameters from 3 to 4 in (75–100 mm), and up to 18 to 24 in (460–600 mm), although the latter may be regarded as exceptional for the conveying of swarf.

Swarf can be fed to the conveyor and discharged at any number of desired points and, in cases where the spiral screw is assembled from paddle or crescent blades, by reversing the inclination of the blades to the axis of the shaft the swarf can be carried in either direction without reversing the rotation of the shaft—see Figure 4.9.

A double-barrelled screw conveyor has been developed having screws of opposite-hand rotating side by side. Conveyors of this type have been found useful in handling bushy swarf as the action of the contra-rotating screws tends to break up the swarf as it progresses through the conveyor.

Reciprocating paddle conveyor

This type of conveyor is very suitable for the transportation of bushy swarf and may well be expected to carry along a considerable proportion of the smaller swarf also. These conveyors are made in two types; in the first the paddles are attached to a captive chain which runs in guides, whilst with the second type the paddles are attached to a 'pusher bar'. The complete mechanism is contained in a steel trough about 24 in wide (600 mm) and the length of stroke is determined by the height to which the swarf must be raised at the discharge end of the conveyor. The method of operation is simple; during the forward action of pushing the swarf along past the tines projecting from the side, the paddles are held rigidly at right angles to the sides and bottom of the trough; on the return stroke the tines hold the swarf which, in turn, swings the paddles through 90° and allows them to pass behind the accumulated swarf. The illustration shows the construction—see Figure 4.10.

A modified form of this type of conveyor is used by some machine-tool designers to clear the swarf from the beds of rotary transfer machines and large bed-type milling machines. On the latter machines, hinged paddles may be affixed to the sides of the table or located under the outer edges of the table, and as the swarf falls from the table it is progressively pushed to the ends of the bed where it falls into removable chip collectors.

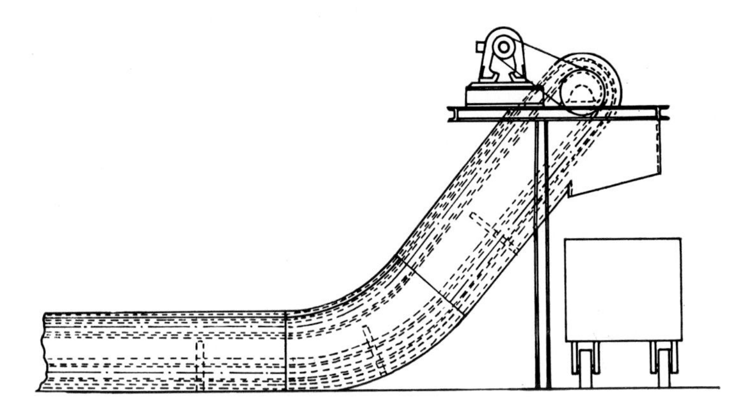

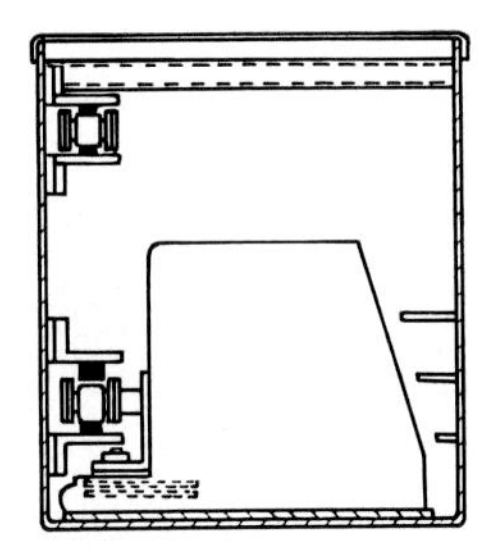

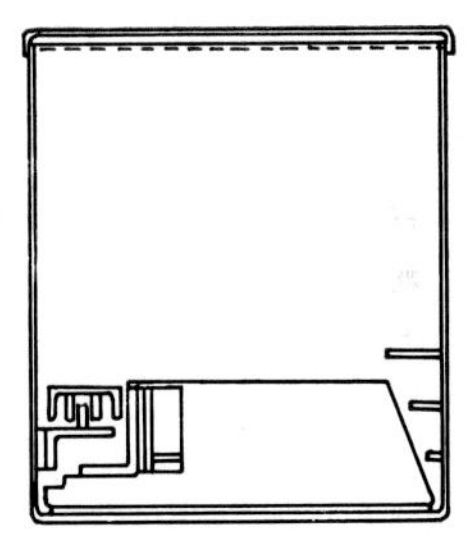

Fig 4.10 A reciprocating-paddle conveyor. The paddle may be operated by chain or by the pusher bar principle

Reciprocating push-bar conveyor

This system is a comparative newcomer to this country but is extensively used in parts of Europe.

The mechanism is of robust construction, functions with complete security and demands the minimum of attention. In principle, the conveyor consists essentially of a tube located in the centre of a U-shaped trough, usually of concrete and let into the workshop floor. This tube is equipped with a continuous slat along the underside which permits it to be supported at suitable intervals from the bottom of the trough. On both sides and the top of the tube, angled fins are fitted and these are the means by which the swarf is moved along. Motion is imparted to the tube by means of a hydraulic ram which has a stroke of approximately 5 ft (150 cm),

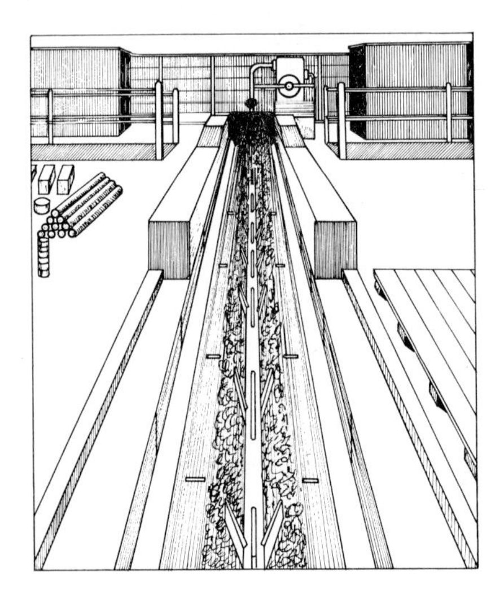

Fig 4.11 A view of a reciprocating push-bar conveyor transporting bushy swarf to a suitable collecting hopper

the speed of travel of the conveyor being 32 ft (975 cm) per minute. The bushy swarf is fed into the trough and the angled fins push it along on the forward stroke past horizontal tines projecting from the sides of the trough. On the return stroke, the tines hold the swarf and prevent it from being drawn backwards. The hydraulic unit consists of a single cylinder having a double-acting piston, and is usually located in a separate pit. The general layout of the system is shown in the accompanying illustration—see Figure 4.11.

5 HYDRAULIC METHODS FOR THE CONVEYING OF SWARF

INTRODUCTION

Hydraulic conveying is the term used to describe the movement of solids by a continuous stream of liquid. When applied in connection with machine tools, solids take the form of swarf in any of its easily recognisable shapes and the liquid is normally cutting fluid. Swarf is washed away from the cutting zone into a trench in which a continuous stream of cutting fluid conveys the swarf hydraulically to a convenient take-off point. The distance over which swarf can be conveyed hydraulically is basically limited only by provision of adequate trenches in the floor and the volume of clean cutting fluid required. By using correctly designed and engineered hydraulic systems it is possible to transport almost all types of swarf, from the finest of grinding or honing sludge to the bushy swarf from heavy machining operations on steel components.

If three conditions on the machine tool are fulfilled, i.e. a copious flow of cutting fluid, jigs that are designed to facilitate the removal of swarf and machine-tool surfaces that will not encourage the sticking or gathering of the swarf, then there will be little difficulty in clearing the cutting area. Having brought the swarf to the base of the machine tool, adequate arrangements must be made for its onward journey to the main transporting system. In the most commonly installed systems, this final conveying usually takes place in trenches which extend downwards into the workshop floor; these are known as velocity trenches and they may be equipped with flushing nozzles.

SOME FACTORS AFFECTING HYDRAULIC CONVEYING

In considering the transportation of solids by liquids, there are two main problems that will need to be overcome. These are:

1 the engineering problems associated with the feeding of the swarf into the main conveying system, pumping capacities and layouts, minimisation of abrasion, etc.;

2 the problems concerned with trench sizes, transport velocities, settling velocities, solids concentration and cutting-fluid viscosities, etc.

One of the latter problems that is particularly important concerns the sizes of the swarf particles that are to be transported and their behaviour during transportation. When the particles are very small, e.g. less than $\frac{1}{16}$ in (1·6 mm) diameter, they will mix with the cutting fluid to the extent that the viscosity of the fluid is changed but the particles do not settle-out. When such a mixture is close to a state of rest it tends to behave as a solid and this means that a minimum shear stress must be exceeded before the mixture will flow. In the case of larger pieces of swarf that come within the second of the classifications given in Chapter 1 and which may be pieces from broken helices, often one or two full turns in length, or chips or slivers, they will mix with the cutting fluid to form a heterogeneous mixture that may be moved by the flow of the fluid in two different ways, these being:

(*a*) in suspension, if the particles are reasonably small and the flow velocity is sufficiently high;
(*b*) by saltation, moving along in a series of intermittent jumps, if the particles are large and the flow velocity is low.

Later in this chapter particular attention is given to the problems associated with the hydraulic conveyance of 'bushy swarf'.

Velocity trenches

The analysis of flow in open trenches is a complex subject because the cross-sectional area is free to change and the velocity distribution is non-symmetrical, the velocity being greatest at the points least affected by the boundaries. Whilst a number of flow formulae has been developed in relation to the transporting of solid/liquid mixtures in pipelines, there is little information available that could be used in designing a functionally reliable system for the conveying of swarf in trenches. Nevertheless, some of the correlations observed during pipeline experiments provide a useful guide to the mechanisms involved in open-trench conveying.

Although both volume and velocity of the conveying fluid are of great importance in transporting swarf in open channels, it must be remembered that turbulence within the moving fluid is the prime factor responsible for moving the swarf.

Trench forms

Without any doubt the success of a hydraulic conveying system depends upon the design of the conveying trenches as the major factor. The most convenient arrangement is that in which the trenches are sunk into the workshop floor with the top of the trench at ground level and a cover plate over. Other methods are used, however, such as with the trenches actually above and resting on the floor (but this causes obstruction to the passage of personnel and transport), raising the trenches to roof level (with the obvious disadvantage that the swarf and cutting fluid must be pumped or otherwise raised to roof level) and it has been known, where there is a basement to the workshop, for the trenches to be suspended

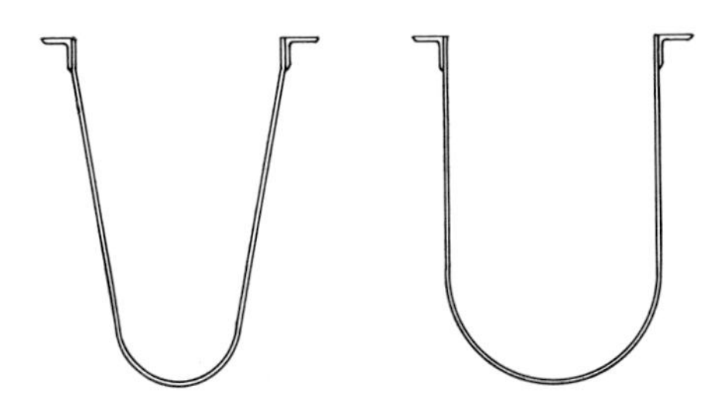

Fig 5.1 Recommended cross-sectional forms for velocity trenches

from the underside of the workshop floor with access holes arranged in the floor for the swarf and cutting fluid to enter the trenches.

The cross section of a velocity trench should always be of 'U' or rounded 'V' form; flat-bottom trenches should never be used—see Figure 5.1.

The layout of trench systems

Trench systems for the conveyance of swarf may be straightforward or complex depending on the application. In the case of a transfer line the system could take the form of a straight run terminating at a settling tank, or to accommodate a series of machines it could take the form shown in Figure 5.2. Whichever

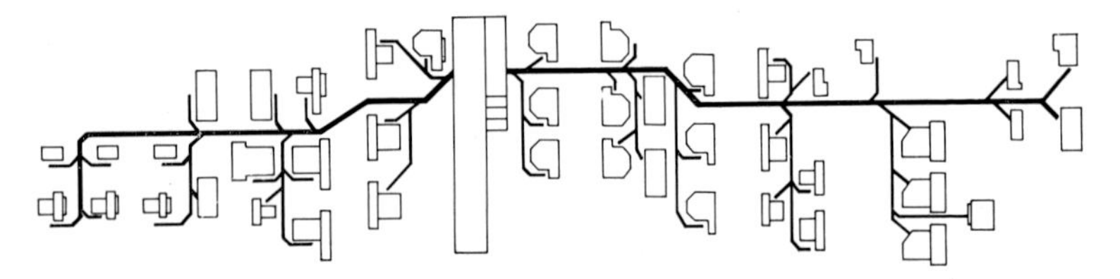

Fig 5.2 Typical layout of velocity trenches for groups of machine tools

it may be, there are certain basic rules that must be observed.

Proposed systems should first be studied with the object of establishing the main trench run into which all other side trenches are to deliver the swarf and cutting fluid. Having decided upon the positioning of the trenches, the slopes must be calculated, taking into account the type of swarf to be transported, the viscosity of the cutting fluid, the length of the trenches, etc., and bearing in mind that all trenches must slope towards the settling tank in the direction of liquid flow.

If the system is designed to rely solely on the action of flushing nozzles to transport the swarf and achieve liquid velocity, then the slope need only be sufficient to make the trenches self-draining. Normally, a slope of 1 in 200 is satisfactory for this purpose, but where there are no flushing nozzles or where flushing is only partly supplying the motive force and the flow of liquid under gravity is supplying the remainder, then the trench slope should be approximately as in Table 5.1.

TABLE 5.1 *Recommended trench slopes when flushing nozzles are not used*

Type of swarf	*Trench slopes*		
		Neat cutting oils	
	Emulsified cutting fluid	*Over 100 SSU**	*Over 300 SSU**
Sludge from grinding, honing and similar operations	1:96	1:77	1:64
Fine chips and heavy grinding sludge	1:64	1:51	1:43
Small-to-medium chips	1:48	1:43	1:38
Bushy and stringy ribbons	1:44	—	—

* Saybolt Second Units measured at 100°F.

Where it is necessary for side trenches to join a main trench, it is essential that the junction should be of the right form and that there should be a waterfall from the side trench into the main trench. The most commonly used junction is one in which the side trench joins the main trench at an angle of 45°. However, it is possible to permit a side trench to enter at angles up to 90°, providing that the main trench has its slope increased for a short distance immediately prior to the junction; this will increase the velocity of flow in the main trench and so prevent a 'back-up' where the side flow enters—see Figure 5.3.

Sometimes, because of the layout of the machine tools or of the need to avoid some static object, it may be necessary to change the

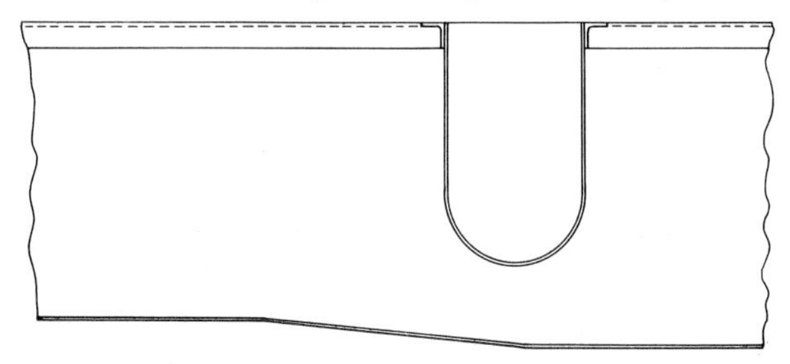

Fig 5.3 Showing the necessary increase in main trench slope if a side trench must enter at 90°

direction of the trench runs. This may be done by using curved lengths of trenching but these are expensive to produce and so it is more usual to use angled sections quite satisfactorily so long as the angles are less than 45°.

When the volume of swarf and cutting fluid is large, it may be necessary to provide a 'spur' in which a flushing nozzle is installed. Sometimes the layout of the machine tools necessitates the entry of two side trenches to the main trench at the same point but from opposite sides and in such cases a spur complete with flushing nozzle should always be provided, particularly so when this occurs at the beginning of the main trench as illustrated in Figure 5.4.

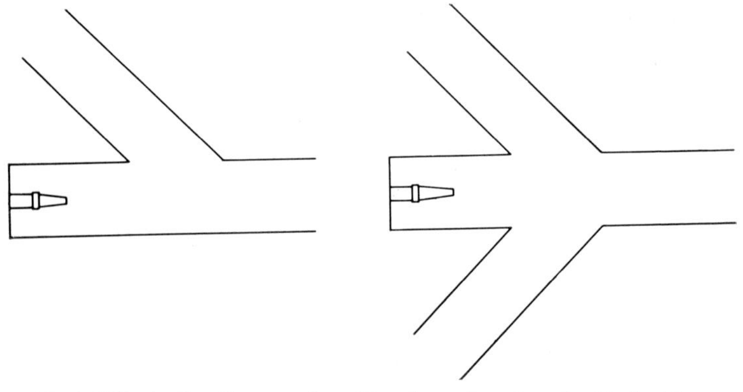

Fig 5.4 Illustrating the use of auxiliary 'spurs' equipped with flushing nozzles, particularly at the beginning of a main trench run

Flushing nozzles

In writing about velocity trenches, mention has several times been made of flushing nozzles and perhaps a few words of explanation are necessary at this point. Flushing nozzles are a means of introducing an auxiliary flow of cutting fluid into the velocity trench; as the name implies, the cutting fluid is injected under pressure, thus increasing the volume and the velocity at the place where the swarf might otherwise be expected to accumulate.

Generally speaking, it is found that flushing nozzles are necessary in all but the simplest systems. To obtain the fullest efficiency the flow rate and pressure to be used at each nozzle must be carefully selected according to the type of swarf to be transported and the general configuration of the velocity-trench system. There can be no hard and fast rules laid down to govern the application of flushing nozzles since the selection is influenced by many factors such as the type of swarf, the viscosity of the cutting fluid, the size of trench and the type of intersection, and it will be realised that the number of permutations covering these factors can be extensive. Engineers who are experienced in the design of hydraulic conveying and central systems will choose the correct nozzle sizes and their most advantageous positioning by using the knowledge they have gained in the field.

Flushing nozzles are usually of the fire-hose type and, to act as a guide, Table 5.2 gives the approximate flows that might be expected when the pressure of the entering cutting fluid is 90 ft head of water at the nozzle (39 lbf in^{-2} (269×10^3 Nm^{-2})).

In any system of trenches in which flushing nozzles are installed, the arrangements should be such that the nozzle may be easily removed for cleaning purposes in the event of its becoming blocked. At the end of a trench where the nozzle is located at the

TABLE 5.2 *Approximate rate of flow of water from a standard nozzle of the fire-hose type when the applied pressure is equivalent to a head of 90 ft (39 lbf in^{-2} ($56{\cdot}5 \times 10^{-4}$ N m^{-2}))*

Gallons (litres) per minute	*Size of orifice. Inches (millimetres)*	*Length of nozzle. Inches (millimetres)*	*Size of pipe connection. Inches (millimetres)*	*Size of flush drop pipe from header. Inches (millimetres)*
20 (90·92)	0·313 (7·937)	4 (101·6)	1·0 (25·4)	1·25 (31·75)
30 (136·38)	0·406 (10·3)	4·75 (120·65)	1·25 (31·75)	1·5 (38·1)
40 (181·84)	0·469 (11·905)	4·75 (120·65)	1·25 (31·75)	1·5 (38·1)
50 (227·3)	0·531 (13·492)	4·75 (120·65)	1·25 (31·75)	2·0 (50·8)
70 (318·12)	0·625 (15·875)	6·75 (171·85)	2·0 (50·8)	2·0 (50·8)
100 (454·6)	0·75 (19·05)	6·75 (171·85)	2·0 (50·8)	2·5 (63·5)

base of the trench this presents no difficulty but, at other points in the system, arrangements must be made for their removal.

All flushing nozzles should have a lock-shield type of valve located in the supply pipe to the nozzle so that the flushing-flow rate may be adjusted to give the best operating conditions.

Several typical examples of nozzle installations are shown in Figures 5.5, 5.6, 5.7 and 5.8.

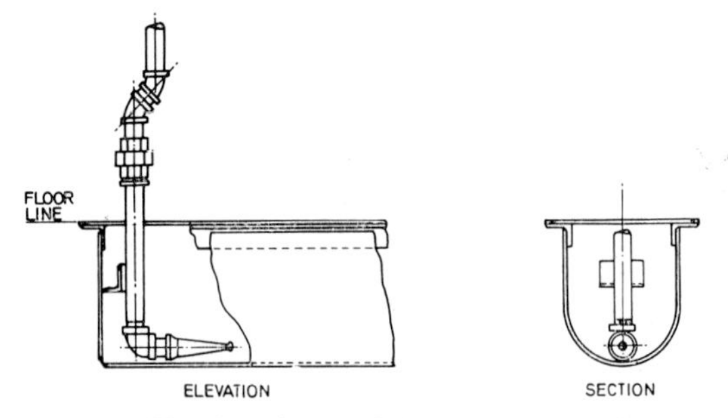

Fig 5.5 Nozzle to be parallel to trench bottom or pointed down to converge with bottom 10′ to 15′ from nozzle

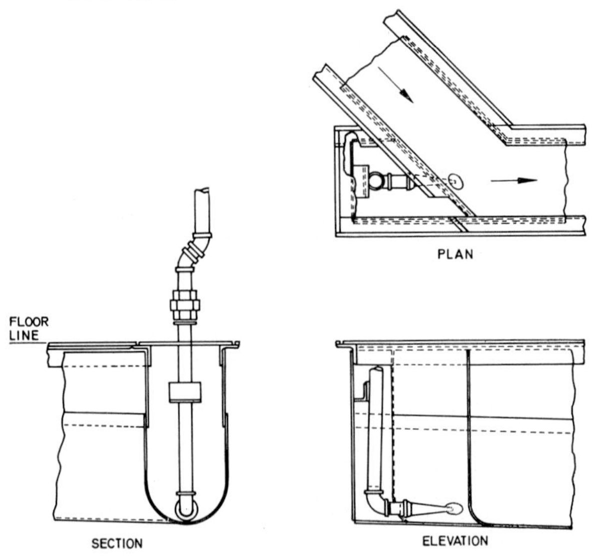

Fig 5.6 Flushing nozzle located in the corner where two trenches meet

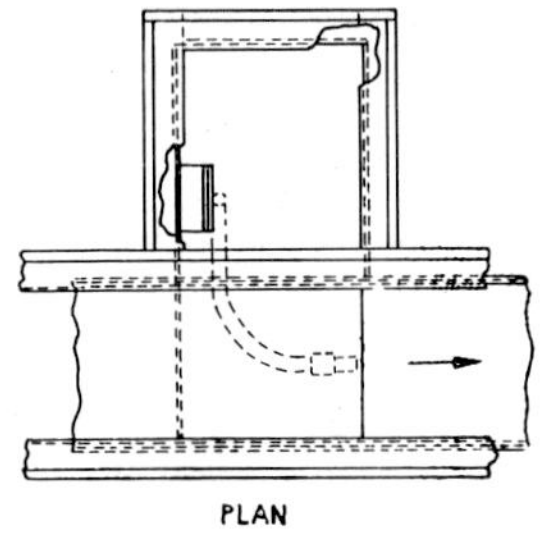

Fig 5.7 Flushing nozzle to increase the velocity in a straight section of trench

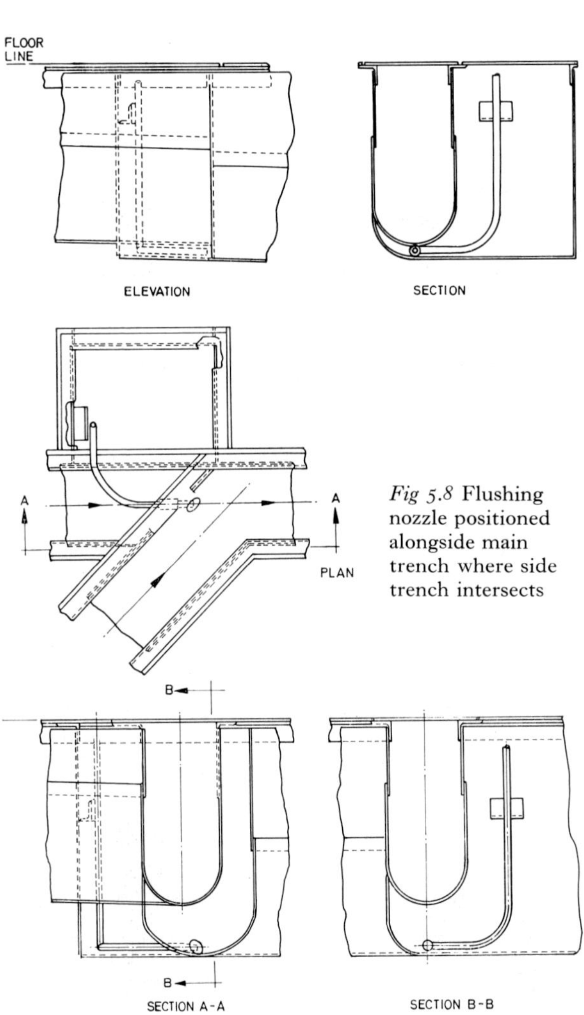

Fig 5.8 Flushing nozzle positioned alongside main trench where side trench intersects

Sizes of velocity trenches

In general, the depth of trenching is established by the following two factors:

1 the starting depth at the shallow end;

2 the slope.

The starting depth should not be less than the width of the trenching and it may be necessary to start with a greater depth if heavy swarf and a large flow of cutting fluid have to be handled.

The width of trenching is also determined by two factors:

1 the flow rate in the trench, where small chips of swarf or grinding dust are to be transported;

2 the type of swarf, where large chips or bushy swarf are to be conveyed.

It is not possible to be specific on the choice of trench widths and experience must usually provide the answer. In Table 5.3 are given trench widths that are suitable for a range of flow rates, but the experience of the design engineer is the most reliable guide in most instances. It is certain that where large chips or bushy swarf are being transported, a trench width of 12 in (30 cm) minimum should be used and widths of 24 in (60 cm) or more may be required where the swarf forms into tangled balls.

It is of importance to realise that, when wide trenches are used for the transportation of the very difficult types of swarf, the flow in the trench must be sufficiently great to carry the swarf regardless of the quantity of cutting fluid being used in the cutting zones. This flow of liquid may exist in the form of a relatively shallow, high-velocity flow with frequent flushing nozzles, or as a somewhat deeper, lower-velocity flow with fewer flushing nozzles but greater trenching slope. If flushing nozzles are used as the primary transporting force then they should be placed at intervals of 20 ft (6 m) along the trenches for bushy swarf and this distance may be increased to 30–40 ft (9–12 m) for less-difficult types of swarf. Again, it is the responsibility of the designer to analyse the situa-

TABLE 5.3 *Rate of flow of cutting fluid in relation to slope and width of conveying trench*

Width of trench. Inches (millimetres)	*Slope of trench*			
	1:96	1:72	1:48	1:36
4·0 (101·6)	0–50 (0–227)	0–80 (0–364)	0–100 (0–455)	0–130 (0–590)
6·0 (152·4)	150–250 (682–1136)	200–300 909–1364)	250–350 (1136–1591)	300–450 (1364–2055)
8·0 (203·2)	350–600 (1591–2728)	400–700 (1818–3182)	500–800 (2273–3637)	800–1000 (3637–4546)
12·0 (304·8)	500–1000 (2273–4546)	700–1500 (3182–6819)	1000–2000 (4546–9092)	1500–2500 (6819–11365)
18·0* (457·2)	1500–3500 (6819–15911)	1800–4000 (8183–18180)	2000–4000 (9092–1818)	2000–5000 (9092–22730)
24·0* (609·6)	2500–5000 (11365–22730)	2500–6000 (11365–27276)	3000–7000 (13638–31822)	3000–10000 (13638–45460)
	Flow in gallon min^{-1} (litre min^{-1})			

* Lower figures are minimum to obtain adequate depth and velocity

tion and to select the best solution, always bearing in mind that, where flushing-nozzle installations that require a step in the trenching are used, then the trenching depth is increasing. Equally it must be remembered that all the fluid used, both on the machines and at the flushing nozzles, must be pumped, stored and filtered and that the fitting of an excessive number of flushing jets or excessively large jets, or both, will add greatly to the total cost of the installation.

An approximate division of flow for an average cutting-fluid system may be taken as—flushing flow $= 1{\cdot}5 - 2 \times$ machine flow—but it must be emphasised that this is only approximate and individual systems may vary greatly.

The problem associated with bushy swarf

Bushy swarf is the most difficult of all swarf to handle, whether by hydraulic or any other method, but a well-designed trench system will convey this type of swarf efficiently and continuously so long as a few basic principles are observed. These are:

1 Steps must be taken to ensure that the swarf can pass freely from the machines and into the trenches. A long ribbon of swarf trailing in the trench and having one end caught in the machine will quickly cause a build-up of swarf in the trench and, if this occurs under the centre of a transfer machine, it can be both difficult and expensive to clear.

2 Trench widths must be adequate to accept the largest ball of swarf that is likely to accumulate and to pass it freely along.

3 Bends in trench runs should be avoided wherever possible but, if they are unavoidable, then a waterfall must be provided and a spur with a powerful flushing nozzle.

4 Trench intersections should be at 45° with a sufficient drop from one trench to the other so that swarf being carried along the main trench is allowed to pass below the bottom of the side trench. Again, flushing nozzles should be used wherever a large volume of cutting fluid and swarf is entering the main trench from a side trench.

Such systems as these tend to be expensive but they have the decided advantage that there are no mechanical or moving parts underneath the machines and so no maintenance is required. Nor is there any mechanism that can suffer damage from bar ends or other foreign bodies that may fall through with the swarf.

Separation of swarf and cutting fluid

With a hydraulic conveying system, the general rule is that two methods are used simultaneously for the separation of the swarf from the cutting fluid, the heavy swarf being removed by a mechanical conveyor and the fine swarf by filtration.

Two types of mechanical conveyor may be used. The design of the main tank that will receive and collect the dirty cutting fluid will normally be such that a drag-type conveyor may be installed and this will continuously scrape along all the swarf that settles to the bottom and carry it up a sloping ramp for discharge into a suitable container. A typical arrangement is shown in Figure 5.9 and this is the only type of conveyor necessary in systems dealing with every kind of swarf from fine dust to small chips. A different type of conveyor may be needed, in addition to the one mentioned above, in systems that deal with swarf in bushy form or as large chips and such a conveyor may take a number of forms depending

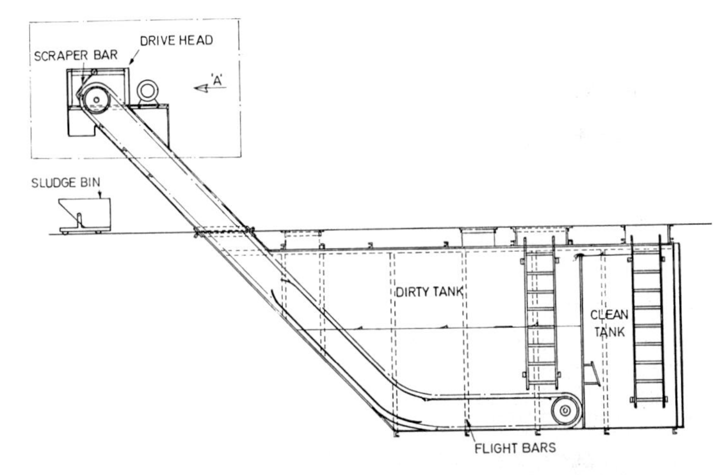

Fig 5.9 Sectional view of a dirty-coolant tank equipped with a drag conveyor and a clean-coolant compartment at one end

on the type of swarf. If the swarf is not stringy nor of the type that forms into large balls, then the banjo-type separator/conveyor may be used—see Figure 5.10. If the swarf is bushy and liable to form tangled masses then the conventional pan-type conveyor is required and this should be mounted inside the dirty cutting-fluid tank below the discharge point of the velocity-trench system, where it will collect the swarf, elevate it and discharge it as required. Alternatively, a reciprocating-bar or paddle-type conveyor may be used but, if this type is used, it must remove the swarf completely without allowing it to enter the dirty tank and it must allow the cutting fluid to drain freely away.

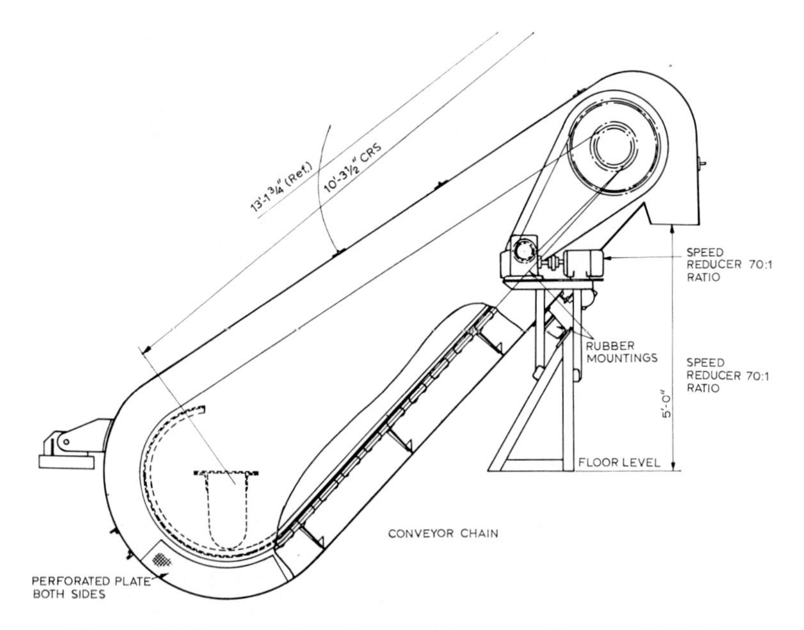

Fig 5.10 Another type of swarf separator/conveyor/elevator suitable for all forms of swarf except 'bushy'

Tank sizes and design

The size of a cutting-fluid tank is controlled by the following factors:

1 The maximum level of the liquid in the tank must always be below the exit level of the velocity trenches during the periods of operation.

2 The capacity must be adequate to provide for the fall in level that will take place when the system is operating and considerable quantities of cutting fluid are circulating in the trenches and the pipelines.

3 The total quantity of cutting fluid must be sufficient to dissipate the heat imparted to the fluid by both the pumping and machining operations.

A general layout of a tank, filter and pump installation is shown in Figures 5.11 and 5.12, the latter being an actual photograph.

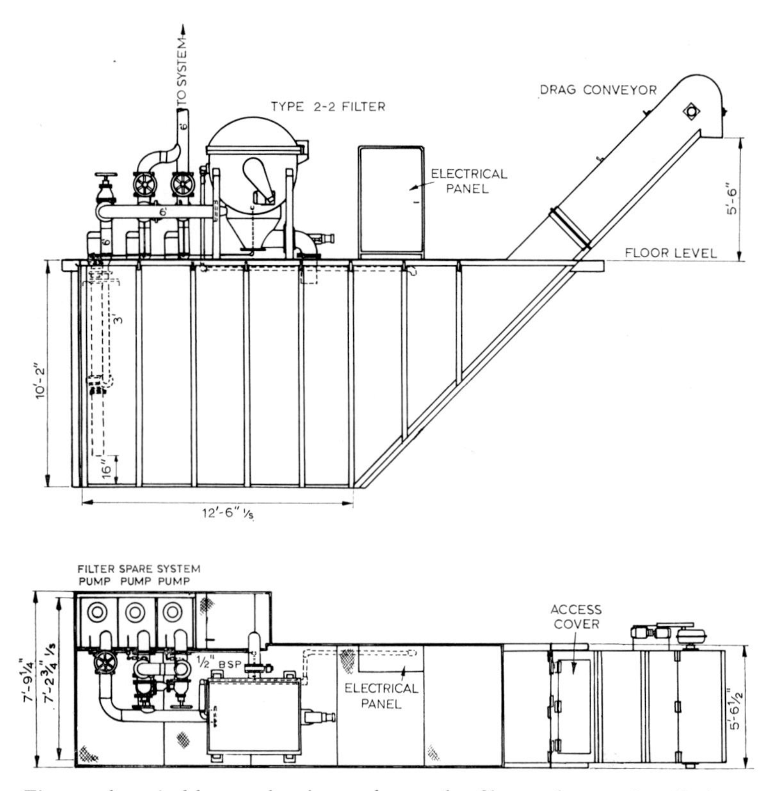

Fig 5.11 A typical layout showing coolant tanks, filter and pump installations

From all that has been written, it will readily be seen that the design of a central system using hydraulic swarf conveying, with a minimum amount of mechanical assistance, can be the most simple method of clearing and conveying swarf from machine tools. The design of hydraulic conveying systems is by no means an exact science, however, and it requires very close collaboration between the engineers and systems designers, and the benefit of all their knowledge and experience, to evolve a system the operation of which can be guaranteed.

Fig 5.12 Photograph of cutting-fluid pumps and filter being installed in a new factory

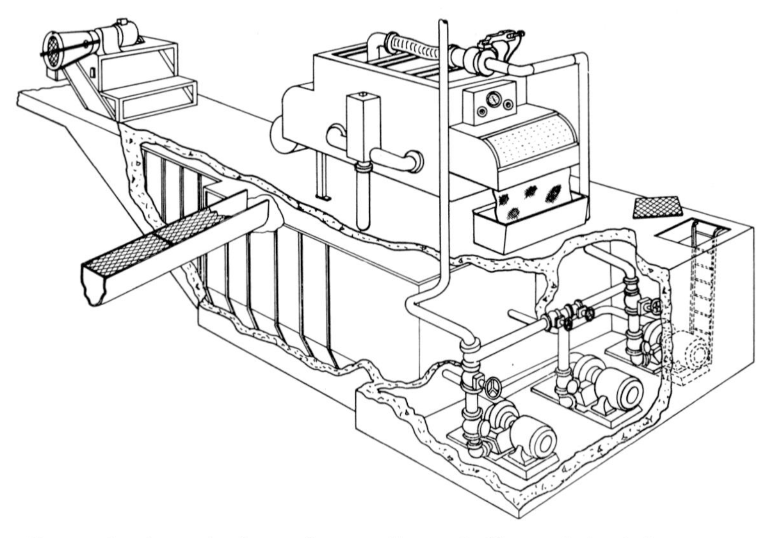

Fig 5.13 An alternative layout for a settling tank, filter and circulating pumps for use in a hydraulic conveying system

6 PNEUMATIC METHODS FOR THE CONVEYING OF SWARF

INTRODUCTION

The scope for pneumatic conveying of materials, metallic or non-metallic, is almost limitless if sufficient time and thought is applied to the subject from first principles. When applied to swarf removal from machine tools the in-built advantage of dust extraction coupled with pneumatic conveyance is a tremendous asset from a health and safety aspect, bearing in mind the machine-tool operator's welfare. Furthermore, with the eradication of fallout of swarf and dust, less maintenance of slideways, bearings, etc., is needed and these advantages must be borne in mind when swarf-removal problems are investigated. The improvement in good housekeeping around the machine tool is enhanced considerably when pneumatic removal of swarf is applied and this is an added bonus that pneumatic conveying systems can offer to a swarf-handling problem.

PNEUMATIC CONVEYING

The primary object of a pneumatic conveying system should be to ensure maximum collection of swarf chips and dust at the point of creation. Generally, swarf and dust extraction is effected by locating a zone of air entrainment as close as possible to the zones of swarf production; the closer the two zones approach, the more effective and economical the plant will be—see Figure 6.1. Usually each machining point will require its own entrainment zone which will be produced by one or more suction hoods attached to the combined swarf and dust extraction unit.

Highly effective chip removal is obtained by appropriate design of the collecting hoods and a rational interaction between an air current and the chip stream near the suction opening. However, uniformity in the behaviour and direction of the stream of chips and dust particles and the aerodynamic peculiarities of the chip depend upon the physico-mechanical properties of the materials being machined, the type of machining and the cutting conditions.

In the cutting zone, continual motion of the air close to the inlet of the collecting hood must be established. Under different cutting conditions the stream of chips and dust will enter the hood at different angles to the leading edge of the cutting tool and, on passing into the area where the air current becomes effective, will be subjected to the influence of three principal forces:

1 the force of gravity;
2 the inertia force;
3 the aerodynamic force.

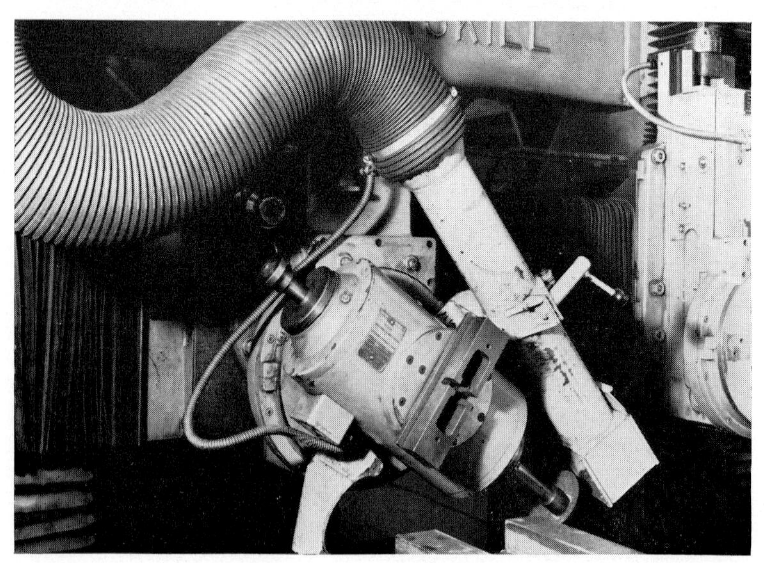

Fig 6.1 Shows how the air-entrainment point is almost in contact with the wheel on a slideway grinder

For instance, when considering the application of pneumatic swarf removal to a milling machine, it must be borne in mind that, under normal cutting conditions, the milling cutter may produce in the collecting hood a circular air current of considerable kinetic energy which will increase with the speed of revolution and the size of the projections on the periphery of the cutter. Under the conditions being considered, large multi-tooth cutters of diameters exceeding six inches and embodying mechanically clamped tool inserts will operate in the manner of a fan. Thus, in the design and calculations for the collecting hood, it is necessary to take into account the effect of such a circular air current on the direction of motion of the particles and to provide means to prevent the chips being carried away from the hood. The velocity of the suction air should be calculated in such a way that the aerodynamic suction forces are greater than the forces that tend to carry the particles away from the collecting hood.

Of prime concern in the design of pneumatic conveying systems, because of its effect on the pressure drop in the system, is the solids content of the solids/air mixture. Generally, the amount of solids handled in low-pressure pneumatic conveying is greater on a weight basis than the quantity of air used.

The usual design procedure is first to select pipe sizes that will provide velocities above the minimum for the material to be conveyed. Next, the pressure drop is calculated on the basis of air alone. This is then corrected according to the material loading.

In order to convey solids properly, the air velocity must always be held above the necessary minimum. The airstream velocity necessary for the continual removal of swarf particles from the collecting hood and for its stable transportation along the ductwork may be taken as the floating velocity, i.e. the mean upward velocity of the airstream in which a single chip will remain in a state of suspension. This support velocity depends on the shape, size and weight of the chip, and it may be determined experimentally or by calculation. Generally, minimum conveying velocities are dependent on a number of factors, including the surface condition of the material and its size, shape and apparent density. It is customary not to make a distinction between the velocities required for vertical and horizontal transport but to design on the basis of a minimum of 5,000 ft per minute (25 m sec^{-1}) for light bulk weight materials to 6,500 ft per minute (33 m sec^{-1}) for the heavier bulk weight materials. The specific velocity for a particular material may be obtained from technical literature but the required velocities are usually determined by experience or common practice. Pressure drops may be calculated by using the curve shown on the chart—Figure 6.2—for both horizontal and vertical flow, calculation being made for the longest run. In order to keep power costs down to a minimum, the lowest practical velocity should be

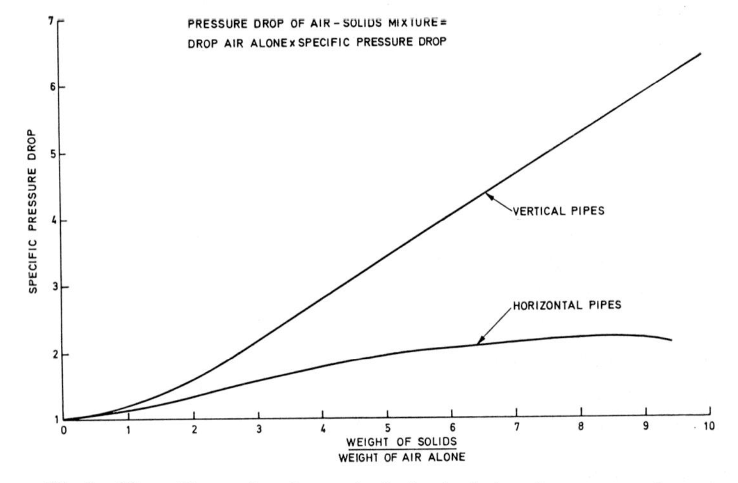

Fig 6.2 Chart illustrating the method of calculating the pressure drop in either horizontal or vertical conveying pipes

selected and the use of long runs of ducting at very high velocities should be discouraged, bearing in mind that the most essential feature is to ensure that the particles being conveyed are not allowed to settle in the conveying ductwork due to underrating conveying velocities, otherwise a build-up in the conveying ductwork will result.

Where swarf is fed from a mechanical conveyor to a suction pick-up boot—a type of hopper with a regulated throat orifice—at a predetermined rate of feed, no extra air movement is required to float the material into the airstream and in this case the design criterion is solely the amount of air necessary to convey a given tonnage of swarf and the conveying velocity itself. It is recommended that the exhaust connection be inclined and provided with an open lower end so that any component or foreign body may fall from the conveying airstream, thus preventing the possibility of a build-up in the conveying ductwork—see Figure 6.3. As an alternative, a venturi type of hopper may be used but this is not generally recommended; if the material to be dealt with is bulky in nature, the efficiency of the venturi tends to be impaired by the turbulence such material will create.

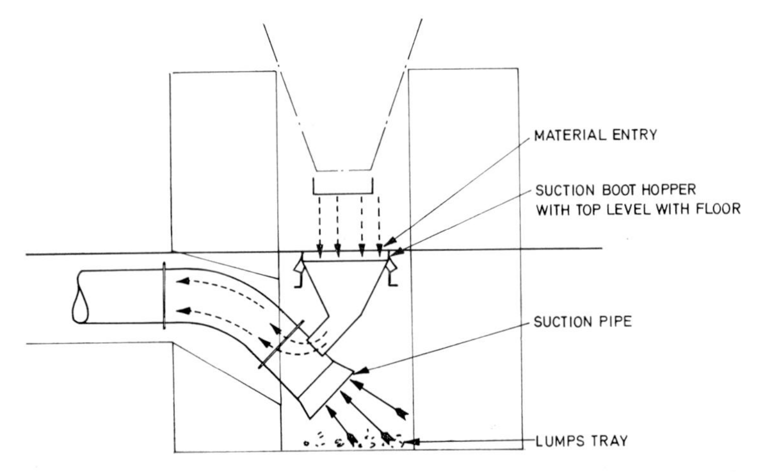

Fig 6.3 Typical suction-boot pick-up arrangement with tramp-iron removal point

In the majority of swarf-conveying applications the amount of air necessary to convey 1 ton (1,000 kg) of swarf per hour ranges from a minimum of 1,500 cu ft (42 m^3) of air per minute to 2,800 cu ft (80 m^3) of air per minute. These figures represent 40–75 cu ft of air per pound (2–4 $m^3\ kg^{-1}$) of material handled per minute.

Cast iron, one of the heaviest swarf materials, can be successfully conveyed at the lower air ratio but a higher ratio will allow a conveying duct of larger diameter to be used offering a decrease in pressure drop and power requirement. In addition, the use of a higher air ratio gives a certain marginal cover for surge loading of material feed rates into the system and also for possible future increases in swarf production from the machine tools.

When, in exceptional circumstances, swarf has to be conveyed from a static pile, considerable air velocity is necessary at the pick-up nozzle in order to cause sufficient air movement to fluidise the swarf and to allow it to become airborne. Although the amount of air necessary to convey a given quantity of swarf will be as indicated in previous paragraphs, at the pick-up nozzle itself velocities must be considerably in excess of the velocities used in the conveying ductwork. Velocities between 50 and 100% above the conveying-duct velocities are necessary for this application and it is essential that nozzles should be so designed that air can be exhausted at all times and that clogging is impossible.

Close proximity hooding

Close-proximity hooding is of prime importance and is the basis of success with most pneumatic-conveying applications. In order to obtain optimum results from a pneumatic extraction system, the design of the close-proximity hoods must be regarded as of paramount importance, the ultimate aim being to enclose the cutting point as closely as possible whilst still allowing clearance for the passage of workpiece and fixture under the cutting tool. The design and location of the hoods should be such that the majority of the swarf produced is directed by the cutting action to the point of maximum air movement. The design must also ensure that the swarf is always airborne from liberation point to exhaust hood so that its path is influenced by the exhaust air velocity throughout its traverse.

In theory, hood design can be related to a few simple principles. The nearer the hood to the source of swarf and dust generation, the less air flow is required and hence the smaller the over-all size of the plant. It may be economical to increase the entrainment velocity up to a maximum of 10,000 ft per minute (50 m sec^{-1}) but the increase in suction required and consequent increase in fan horsepower must be weighed in the final economic evaluation.

The need for the hoods to be designed and located so as to enclose the cutting point as closely as possible to ensure that the maximum quantity of swarf is extracted is clearly explained by reference to the graph in Figure 6.4. This graph shows the velocity contours and streamlines for air flow into a circular orifice; as an instance, it is shown that only 5 in (127 mm) away from the orifice the air velocity has fallen to 20% of the inlet velocity. Once a particle of swarf is liberated by the cutting tool and entrained in the airstream, it is almost impossible to predict the path it will follow due to collision with other particles, etc., but

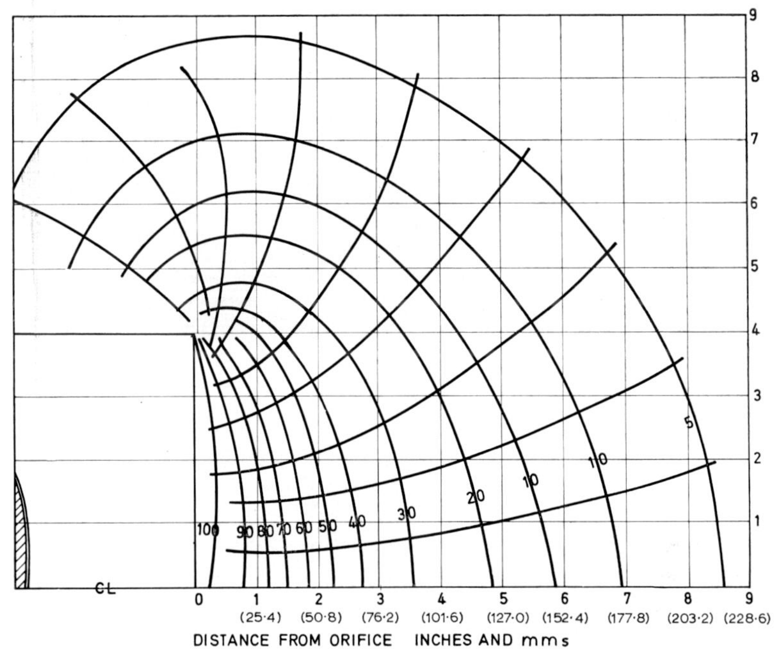

Fig 6.4 Velocity contours and streamlines for circular orifices. The contours are expressed as percentages of the opening velocity

with machining operations such as the milling operation shown in Figure 6.5, the application of close-proximity hoods makes it possible to obtain extraction efficiencies of 96–98% by using air velocities well in excess of the cutting speeds.

It should always be remembered that provision must be made for air to enter the entrainment zone freely as restriction here will starve the main plant and result in the choking of connections, deterioration in swarf and dust control, or increased fan horse-

Fig 6.5 A close-proximity hood used with a face-milling cutter

power to overcome the intake restrictions imposed by close-proximity hoods.

Once the swarf is entrained in the airstream within the exhaust duct, the actual air velocity can be reduced to a lower level, since less air is required to convey a given quantity of material than is needed to collect it. On close-proximity hooding systems, a duct-work velocity between 5,000–6,000 ft per minute (25–35 m sec^{-1}) is the normal application for most metals. However, careful consideration must be given to individual materials and a minimum transport velocity for any material to be conveyed can quickly be determined by reference to a Madison conveying chart, an example of which is shown in Figure 6.6; the velocities given include sufficient margin to avoid blockage in systems. As a general guide, materials with bulk densities ranging between 80–100 lb per cu ft (1,300–1,600 kg m^{-3}) can be transported satisfactorily with a velocity of 6,500 ft per minute (33 m sec^{-1}).

Close co-operation with the machine-tool designer is essential right from the concept of original design, so that close-proximity hooding of cutting tools can be accommodated without interference to cutting methods and materials and to handling requirements. Usually the cutting tool must move spatially relative to the work, or vice versa. This often calls for the hood, locating the zone of entrainment, to have the ability to move in sympathy. Careful design and fitting is required to achieve this condition if the resultant hood and connection are to give effective service for an

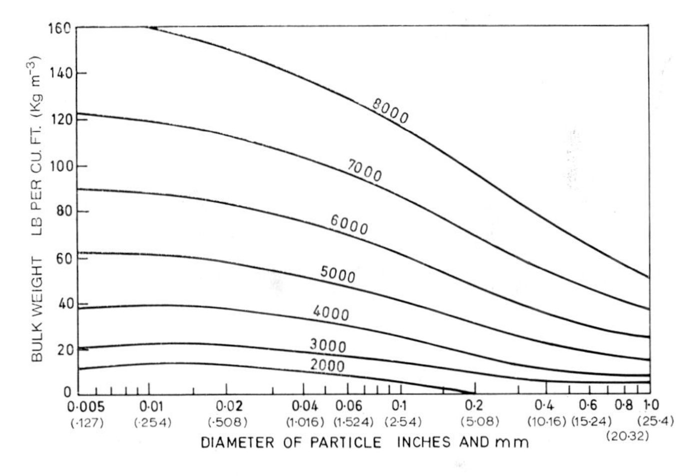

Fig 6.6 A Madison chart showing the minimum transport velocities required for the pneumatic conveying of solids

acceptable period of time. Facility for hood movement is usually obtained by the use of flexible pipe connections, slides, belt ducts, telescopic joints, universal swivel joints, pivots or other suitable methods. It should be remembered that this equipment will be subject to the passage of swarf and dust at high velocity, possibly resulting in wear through abrasion, or deterioration through the effect of increasing temperature. Flexible pipes are particularly prone to failure from this aspect, often accentuated by excessive bending and small-throat radii at the extreme limits of movement.

The operation of a cutting tool usually produces a particular direction of flow of material. Pre-knowledge of this behaviour is extremely valuable, because the ability of the air entrainment zone to change the path of moving particles is far more limited than is normally assumed to be the case, and the larger the particle the more difficult it becomes to change its velocity. Consequently, it is desirable to locate the hood to take maximum advantage of the initial flow path of the material—see Figure 6.7.

Fig 6.7 Multi-extraction hoods applied to a milling machine. Note the utilisation of nylon bristle barriers to direct the swarf into the hoods

Optimum hood design can be obtained only by meticulous and painstaking attention to detail. It is a matter of regret that the experience and time required to achieve good hood design on complex applications when swarf and dust load, tool movement, space restrictions and access requirements all impose restrictions on hood and connection concept, are in limited supply on the one hand, and difficult to reflect in the sales price of the plant on the other. Nevertheless, the greatest possible degree of harmony and understanding between designer and user of hood designs is worthy of any effort to achieve this understanding.

To prove conclusively the effectiveness of a proposed hooding design on any plant of size, it is often a good investment to carry out preliminary trials using a light-gauge hood allied with a simple source of air entrainment and collection; very often a dust collector will answer this purpose. Limited, but not too limited, operation of such a system enables a full assessment of hooding effectiveness to be ascertained as well as providing accurate data relative to swarf and dust extraction, load and behaviour.

TYPES OF MATERIAL TO BE HANDLED

(a) Cast iron

This material is in most cases suitable for pneumatic conveying without the necessity for further processing. Systems of any size can be designed to deal with any type of machine tool ranging from simple single-operation machines up to the larger multi-station automatic-transfer line.

Close-proximity hooding around the tool is highly desirable as, in order to ensure first-class working conditions for the machine-tool operator, it is almost always necessary to exhaust the fine dust which is produced with cast-iron swarf.

Where it is not possible to apply close-proximity hooding and the swarf is allowed to fall to the machine bed, it is preferable to consider the use of a drag conveyor to collect and deliver the swarf to a central pick-up suction point, rather than attempt to exhaust the swarf pneumatically from small hoppers in the machine bed. The inevitable leakage from machine hydraulic systems can cause acute bridging problems if the latter method is attempted—see Figures 6.8 and 6.9.

(b) Steel

This material is not suitable for pneumatic conveyance at source unless effective chipbreakers are fitted and maintained in use. Bulk conveying of steel swarf from mechanical conveyors is also impracticable unless the material is first broken down to a suitable size.

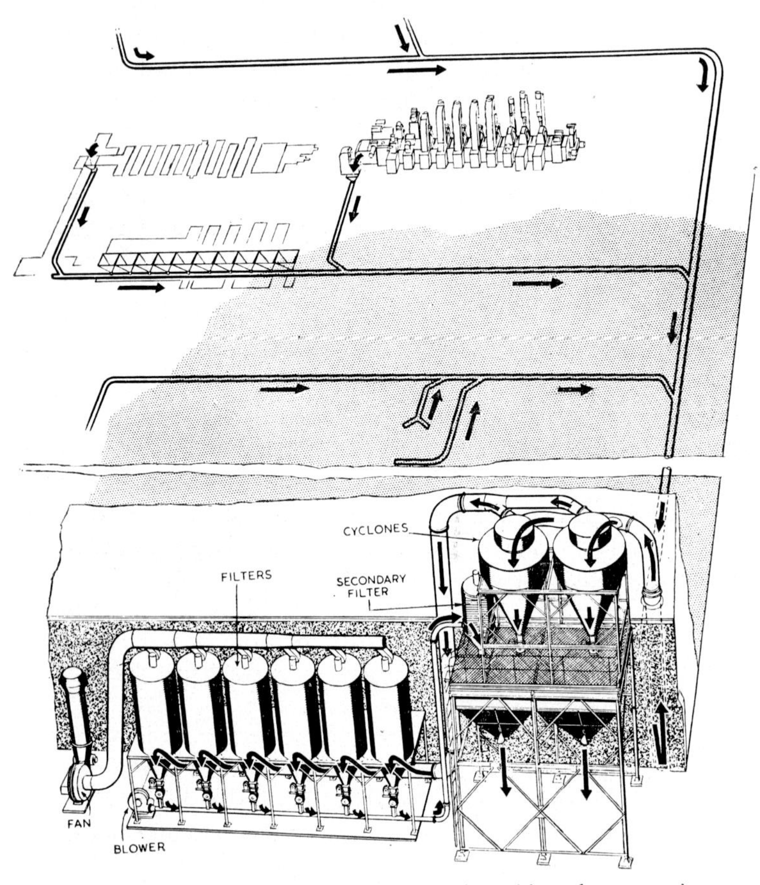

Fig 6.8 Layout of central pick-up suction station with cyclones as primary collectors and fully-automatic fabric filters as secondary collectors

As a general guide, steel swarf can be regarded as unacceptable to a pneumatic conveying system due to size and shape of chips produced and in almost every case the presence of moisture due to cutting fluids being used in the cutting operation.

(c) Aluminium and light alloys

Generally this material presents no serious problems and it can be handled in an identical way to cast iron, bearing in mind its lower

Fig 6.9 View of cyclones and fully-automatic fabric filters. Note large collecting hoppers and loading points

density and the possible reduction in conveying velocities. Occasionally, with different cutting methods, fine particles are produced and this problem must be carefully borne in mind when a suitable collector is being considered.

Large curly chips can often be found at the cutting and delivery point but, with the correct application of close-proximity hooding with internal baffles, breakdown of particle size can quite readily be obtained to enable successful conveyance.

A most important aspect with regard to aluminium and other alloys is the incidence of fire and explosion hazard. Factory regulations controlling the handling of swarf or dust from magnesium, aluminium and light alloys provide clear directions on plant design when these materials are involved.

(d) Bronze, brass and copper

The swarf from these materials tends to curl in small sections and close hooding of the operation is nearly always possible. Careful consideration must be given to the possibility of moisture content due to coolants being used in the cutting operation; a limit of 6% maximum moisture content must be strictly adhered to if trouble-free systems are to result. If the moisture content is in excess of 6%, preliminary processing of the swarf via a centrifuge is essential if pneumatic conveyance is to be considered practicable.

(e) Plastics

Small swarf and dust particles are produced when machining these materials and, due to the wide variety of materials available, there is considerable variation in the nature of the swarf produced.

Phenol-based plastics produce dust in great quantities and close-proximity hooding is necessary. Static electricity can be a problem both at the collection point and at other points in the system, and adequate earthing material is essential to reduce the explosion hazards which are possible in certain cases.

Resin-bonded materials must be carefully considered due to the tacky nature of the swarf produced and the readiness to build up on internal surfaces of conveying ductwork.

(f) Graphite and carbon

This material is readily conveyed as it is dry and granular in nature. The swarf given off is easily handled and close-proximity hooding is strongly advised in all cases because of the high percentage of fine dust produced by the cutting operation.

(g) Asbestos

The machining of asbestos-based products produces dust as well as swarf and, because of the health hazard involved, close-proximity hooding is absolutely essential to comply with factory regulations governing the machining.

This material is very suitable for pneumatic conveyance but careful consideration of collection equipment must be of prime importance in the design stage of any proposed system

FACTORS AFFECTING THE DESIGN OF A PNEUMATIC CONVEYING SYSTEM

In order to design a pneumatic conveying system, full details of the amount and nature of the material to be conveyed must be made available to the design engineer. The over-all path which the material must follow, from pick-up to delivery, must be given careful consideration and in the interests of good design and economics the designer should eliminate any unnecessary bends in the ductwork which conveys the swarf from cutting-point liberation to final collection zone. The conveying distance should also be kept as short as possible.

If large quantities of swarf are produced, difficulty may be experienced in ensuring a regular flow of material to the pneumatic conveying system. For instance, if the swarf is allowed to come to rest on a base plate or any other machine-table components, it is exceptionally difficult to introduce it into the pneumatic system solely by the flow of air, mainly because air always takes the least line of resistance. Once airborne, however, few difficulties are involved in the conveyance of the material, and the method of removal or disposal from the air path has then to be considered carefully and is dependent entirely upon the size, shape and weight of the swarf. If continuous ribbons of swarf as distinct from chips are involved, then steps must be taken to break them into more convenient lengths by means of crushing equipment which can be installed at the inlet to the suction points. Alternatively, baffle-plates can be fitted in the exhaust hood so as to alter the direction of flow of material, thereby reducing its size by impact breakdown. Crushing equipment can also be utilised separately to accept the swarf delivered from the machine or transfer line via a mechanical or vibratory conveyor and to deposit the broken swarf into the inlet of a pick-up boot of a pneumatic conveying system.

In addition to the nature of the swarf being handled, the question of moisture content, particularly of oily material, is of great importance. Whilst the use of a liquid lubricant or cutting fluid may be incidentally beneficial in reducing the generation of free, airborne dust, it imposes many complications on a combined swarf and dust collecting unit, from the need to make the whole extraction system liquid-tight to imposing severe restrictions on the selection of collector. Machine-tool cutting lubricants, whether soluble or mineral oils, also create problems for the pneumatic-conveying engineer due to the fact that, once in transmission in the conveying ductwork, the centrifugal action of conveyance tends to deposit the smaller particles of oil and dust on to the walls of the conveying ductwork, with a resultant increase in frictional resistance to air flow which can eventually lead to a blockage of ductwork and ultimate failure of the plant. A possible remedy in certain circumstances would be to arrange for drying equipment, i.e. infra-red drying panels, to be installed before the swarf is delivered to the pneumatic-conveying pick-up boot whenever moisture is present with the swarf to be handled.

For general information, the following points result from past experience and in some cases will apply equally to close-proximity hooding applications as well as bulk conveying systems:

(*a*) Avoid the use of fabric filters with pneumatic conveying systems handling wetted swarf.

(*b*) Do not attempt to convey bushy or curly swarf (i.e. steel or brass) pneumatically from a static pile.

(*c*) Make all collection-hopper outlets large enough to avoid bridging. This also applies to open-top hoppers under machines, which should allow the full quantity of swarf to enter by gravity without the need for assistance from the suction exhaust connection.

(*d*) Do not use open-top hoppers under machines unless there is adequate access for maintenance.

(*e*) Ensure that all ductwork layouts are as direct as possible using the minimum number of easy-radius bends. Bends having a throat radius less than 3 pipe diameters should not be fitted.

(*f*) Ensure that adequate inspection doors are fitted to all ductwork, at frequent intervals on horizontal duct runs, at all bends and offsets and branch duct-shoe connections. All ductwork to have flanged joints and capped rodding points when the moisture content of the swarf being handled is approaching the limit of 6%.

(*g*) All conveying ductwork to be airtight on erection to avoid air losses at joints, etc.

(*h*) One of the largest single factors influencing the capital cost of a pneumatic conveying system is the thickness of ductwork utilised.

For non-ferrous swarf the minimum wall thickness of ductwork should be $\frac{1}{8}$ in (3 mm) plate and for ferrous metals at least $\frac{1}{4}$ in (6 mm). Mild-steel plate should be utilised when conveying in bulk.

At all right-angle bends the duct thickness should be increased by 100% and, where necessary, such bends should be fitted with replaceable wear plates on the outer scroll where the greatest wear is anticipated.

All pneumatic conveying applications should be very carefully examined for abrasive properties of swarf at the initial design stage of any scheme and one of the largest single factors influencing the capital cost of a pneumatic conveying system is the thickness of the ductwork utilised. Recommended thicknesses of ducting material are given in the section dealing with the design of conveying ducts.

(*i*) Avoid passing bulky swarf through the exhaust fan. Crushed swarf in low concentrations can be handled successfully by a paddle-blade fan but, to avoid wear, out-of-balance characteristics and to keep motor horsepower to a low level, it is always sensible to install the exhaust fan on the clean-air side of any swarf separator.

(*j*) Where continuous delivery of swarf is required from the high-efficiency cyclone collection hopper, a rotary valve of oversize capacity should be installed; slow running and the use of neoprene or similar tips fitted to the rotor impeller are essentials for trouble-free installations.

When considering the installation of a pneumatic system, it is always wise to secure the co-operation of the machine-tool operators from the outset. Whilst they will appreciate the cleaner working conditions, these must not interfere with their earning potential and attention to the following points must be given:

1 The design of hoods and connections should be such that a minimum of maintenance is necessary, thus reducing machine downtime.

2 The hoods and associated equipment should provide minimum obstruction to the normal operating of the machine.

3 The operator must always have visual access to the cutting tool and manipulation should not be made more complicated.

MOTIVATION AND COLLECTION EQUIPMENT USED IN PNEUMATIC CONVEYING SYSTEMS

(a) Motivation

The most usual method of creating the necessary air movement is by use of a centrifugal fan. The air flow-rate/pressure-drop characteristic required will tend to dictate the type of fan, and advice on this matter can be obtained readily from the specialist manufacturers. Basically, two types of fan may be considered—the common paddle-blade type and the aerofoil-section, high-efficiency fan.

The centrifugal paddle-blade type of fan is very robust and reliable and a minimum amount of maintenance is required. These fans have been used for dust-extraction purposes for very many years and are simple in operation. During the past few years, however, the aerofoil-section impeller type of centrifugal fan has become popular for use with pneumatic systems. It has higher efficiency than the paddle-blade type and the distinct advantage of requiring less horsepower. However, because the out-of-balance characteristic is much more critical with this type of fan, regular and efficient maintenance is necessary.

The position of the fan in the system is important. With both fabric and wet collector units the fan is normally on the clean side which is advantageous for many reasons:

1 There is little or no abrasion of the fan impeller because there is relatively little dust passing through.

2 This positioning allows highly efficient aerofoil fans to be used, which offer a significant saving in running costs.

3 The collecting unit is operating under suction conditions and any leakage in the system will entrain air rather than blow dust-laden air into the workshop.

4 The positioning should be such that access for inspection and

maintenance purposes is easy; it is assumed that the design of fan selected will have the same advantages.

5 All fans make noise and this is an important point to bear in mind during selection; it may well call for silencing arrangements if the fan is situated near to operators.

The driving motor for the fan must be selected with the assistance of an experienced manufacturer in view of the load it must carry for long periods, and the associated electrical control equipment must be located where it is easily accessible and free from swarf and cutting fluid.

(b) *Collection equipment*

The selection of the correct collection equipment will depend on:

1 the performance rating, either continuous or intermittent;
2 the degree of collection efficiency required;
3 the swarf and dust burden of the air and their characteristics;
4 the method of storage and disposal of the collected swarf.

The following factors are pertinent to the selection of all collection equipment:

(*a*) The maximum period of uninterrupted operation affects the specification of the collector more than anything else. If the machine produces dust and swarf for long periods without interruption, the dust and swarf control system must be continuously rated.

(*b*) Obviously, the nature of the material being machined will affect the design of plant, with emphasis on the hazards of fire and explosion. Home Office regulations are stringent where the handling of magnesium or aluminium dust is concerned and lay down clear directions on plant design; this also applies where many toxic materials are concerned. It is essential that the collector shall continue to function effectively throughout the time and range of swarf and dust concentrations handled, and that the method of collected swarf and dust disposal is effective throughout.

Expansion chambers

Where a large volume of swarf particles of relatively large size with little or no dust content is being handled, expansion chambers are often used. Such swarf will readily be separated out of the conveying stream by change of direction, rapid reduction in velocity and the normal effect of gravity when led to an expansion chamber of sufficient size and adequate capacity to handle the volume of swarf being passed through.

To prevent an effluent nuisance due to any fine dust present, it is essential to fit some form of secondary collector following an expansion chamber.

Cyclones

As an alternative to expansion chambers, a cyclone may be used. This equipment has a greater collection efficiency than the expansion chamber as may be seen from the curve relating particle size with cyclone efficiency shown in Figure 6.10. High-efficiency cyclones can also be used as main collectors for materials such as

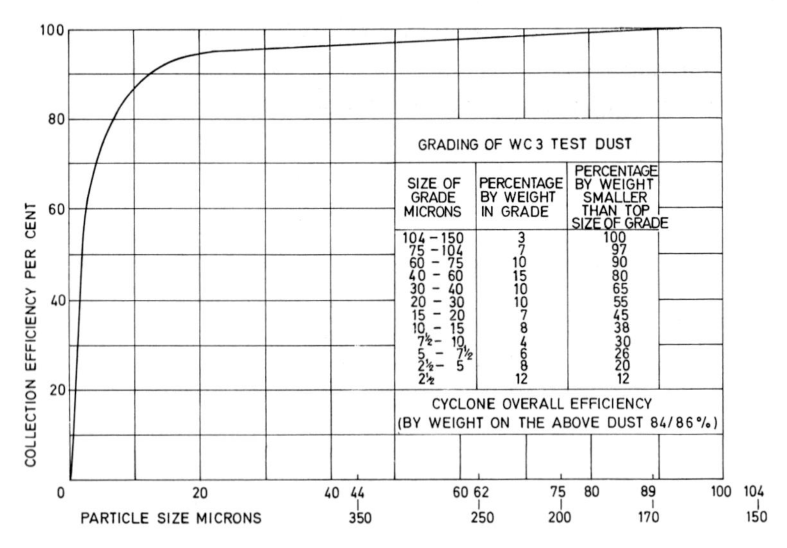

GRADING OF WC 3 TEST DUST

SIZE OF GRADE MICRONS	PERCENTAGE BY WEIGHT IN GRADE	PERCENTAGE BY WEIGHT SMALLER THAN TOP SIZE OF GRADE
104 – 150	3	100
75 – 104	7	97
60 – 75	10	90
40 – 60	15	80
30 – 40	10	65
20 – 30	10	55
15 – 20	7	45
10 – 15	8	38
7½ – 10	4	30
5 – 7½	6	26
2½ – 5	8	20
2½	12	12

Fig 6.10 Grade-efficiency curve for high-efficiency cyclones

non-ferrous metals where there may be a high quantity of fine dust particles in the swarf being extracted—see Figure 6.11.

Although it is usually possible to locate cyclones so that the discharged air is not an embarrassment to the surrounding area or near neighbours, it is essential that expert advice is sought because of the serious consequences that can result when fine

Fig 6.11 Large graphite swarf and dust extraction plant incorporating high-efficiency cyclones

dust is emitted to atmosphere. With the advent of the Clean Air Act, people have become conscious of emissions that are visible either on the equipment or surrounding buildings and this has created a problem that can be solved only by efficient sealing and almost 100% collection. To overcome this difficulty it is recommended that the medium- or high-efficiency cyclone be regarded as the primary collector with a secondary collector, as illustrated in Figure 6.12, to extract the ultra-fine particles before the air is discharged to atmosphere.

The cyclone has been used as a separating medium for very many years. The separating effect depends upon the creation of a

Fig 6.12 High-throughput, high-efficiency cyclone with 'wrapped-around' inlet, handling 5 tons (5,080 kgm) per hour of non-ferrous swarf

vortex when the mixture is introduced tangentially into the tapered cyclone chamber together with an acceleration that may reach a maximum of four hundred times that due to gravity, the angular velocity at the tapered walls producing a hyperbolic increase in the velocity.

As the air/swarf mixture enters the cyclone, the heaviest particles are thrown against the walls where friction robs them of their velocity and allows them to fall to the swarf outlet. The forced vortex has not enough centrifugal force to throw out the finer particles; these are carried down until they enter the free central vortex whose far greater energy throws them back into and through the forced vortex until they too strike the walls and finally fall to the outlet. The cleaned air from the free vortex escapes through the thimble at the top of the cyclone.

With swarf-extraction systems handling fairly large swarf particles a cyclone will be utilised as the main collecting medium; however, where fine dust and swarf from aluminium, magnesium and titanium are handled, a wet collector must be utilised to conform with the Factory Act regulations for those metals which are potentially explosive when in a finely divided state. The provisions of the Factory Acts are explicit in regard to the control of any type of dust which is liable to explode on ignition and the dust created during the machining of the three materials just mentioned falls into this category. One source of ignition may be 'sparking' caused by the swarf particles or other metallic foreign bodies making violent contact with the walls of the cyclone. As the Acts call for the exclusion of all possible sources of ignition the use of cyclones should be prohibited under these conditions.

Wet collectors

There are many designs of wet collector available, the most common type being the self-induced-spray-type wet collector which has the virtue of giving a constant collection efficiency independent of size and having modest floor requirements for any capacity—see Figure 6.13.

Fig 6.13 10,000 c.f.m. (283 m^3) wet collector handling aluminium swarf dust, fitted with explosion door and acoustic silencers on the twin fan outlets

The collection efficiency of self-induced-spray-type collectors is higher than that obtained with high-efficiency cyclones, albeit their energy requirement will also be higher as will their maintenance costs. For most self-induced-spray-type wet collectors a pressure drop of 6-7 in (150-175 mm) water gauge is created across the collector in order to obtain maximum collection efficiency. By comparison with fabric filters, these wet collectors will give an efficiency of 97% when tested with W.C.3 test dust. (See Figure 6.10 for particle sizes and percentages by weight which are contained in what is known as W.C.3 test dust.)

The main advantage of a self-induced-spray-type wet collector is that the only working medium is the centrifugal fan situated on the clean side of the collector. This directs the dust-laden air into the inlet of the wet collector where it is projected immediately on to the surface of the water; an abrupt change in direction takes place and the heavier particles of dust are arrested at this point without causing abrasion by contact with metallic surfaces. The dust-laden air is then pulled through a high-velocity impact zone where a tremendous turbulence takes place and the main de-dusting action occurs. Following the impact zone, metal baffles are designed to create a spray area, again to create a third stage of de-dusting, after which velocities are very rapidly reduced to prevent any water carryover being exhausted by the fan.

The possible effect of the interaction of dust and water in conditions of brisk agitation requires careful examination. This is usually innocuous but may lead to foaming, frothing or retarded settling. If a wet collector must be used under these conditions there is usually a chemical treatment available capable of solving the problem but this may involve regular attention to the water, an added complication when simplicity should be the aim in most swarf and dust control systems.

The swarf and dust is collected in the form of slurry or sludge which can be removed mechanically from the wet collector by a drag-link conveyor. This method of emptying calls for planned maintenance as the conveyor has a finite capacity for sludge removal. Loads in excess of this capacity inevitably lead to mechanical breakdown and the unpleasant task of manually cleaning the collector. Other methods of emptying are available such as manual discharge from high-level hoppers of large capacity, or continuous irrigation with an external settling pond and re-circulation tanks, although the scope for the latter system is usually limited.

Viscous-type dust collector

This type of collecting unit is suitable in applications where relatively fine particles of dust are present, of the type that is produced when cast iron is machined. However, it is limited to dealing with small quantities of entrained dust and, as a consequence of this, a cyclone pre-cleaner is normally used to extract the large swarf particles.

The operating principle is simple and effective, the dust being separated from the airstream as it passes through an oil-irrigated, crimped-wire, fluid-bed pad. The dust-laden air is mixed with a non-detergent oil at a perforated distribution plate over which the oil flows; air passing upwards through the plate entrains oil droplets and deposits them upon the viscous pads which are mounted in the form of an inverted 'V' and are, therefore, self-cleaning. The dirty oil drains from the pads into gutters and then returns to the settling tank where the dust particles are able to 'settle-out', the clarified oil being re-circulated to the distribution plate. The collected swarf material is raked from the tank weekly as a dense sludge and flushing of the viscous pads is necessary every six to eight weeks. Flushing of the pads involves pouring cutting fluid on to their top surfaces and allowing this to drain into the settling tank, the unit being shut down during this process. The addition of cutting fluid to the normal oil is not detrimental to the functioning of the unit up to a dilution of 50%, corresponding to a full year of continuous running.

Such collectors have been widely used during machining operations on cast iron where, although swarf removal has not been the prime object, large quantities of swarf have been removed along with the volumes of fine dust it was desired to collect.

Fabric filters

Fabric filters offer the best solution to the collection of dust in swarf-handling schemes, chiefly because of their high collection efficiency. Many designs of filter are readily available, intermittently or continuously rated. Nearly all now operate under suction, discharging the dust through a mechanical seal or, in the case of small dust loads, storing it in bins placed underneath the filter hopper until the end of the shift, when the plant can be stopped and opened to atmospheric pressure conditions—see Figure 6.14.

The main criterion for successful fabric-filter operation is the correct selection of filtration velocity. Generally, with woven fabrics, this will vary from 1 to 5 ft (30 to 150 cm) per minute; with felted fabrics, higher velocities are used in the range of 5 to 15 ft (150 to 450 cm) per minute.

As a general guide, with woven-fabric filter installations the following filtration rates should be adhered to but, with felted-fabric filters, filtration velocities can be increased by 2 to 3 times with safety.

Spheroidal graphite iron, graphite and carbon,
1 to 2 ft (30 to 60 cm) per minute filtration velocity.

Cast iron and plastics,
3 to 4 ft (90 to 120 cm) per minute filtration velocity.

Non-ferrous metals, aluminium and alloys,
4 to 5 ft (122 to 152·4 cm) per minute filtration velocity.

Asbestos,
2 to 3 ft (60 to 90 cm) per minute filtration velocity.

Fabric filters must be cleaned periodically to remove the dust caked upon the filter elements, this being necessary to overcome the build-up of filter resistance. With a fan-exhaust system such a build-up can be prohibitive as, with any increase in circuit resistance, the air volume is decreased and a fall-off in extraction and collection of swarf results. Filter cleaning can be achieved by mechanical agitation of the filter elements with or without reverse-air cleaning.

With carefully selected filtration velocities, correct application of fabric filter, and an efficiently designed cleaning mechanism applied under suitable conditions, collection efficiencies of virtually 100% can be obtained.

Many commercial designs of all three types of collectors are

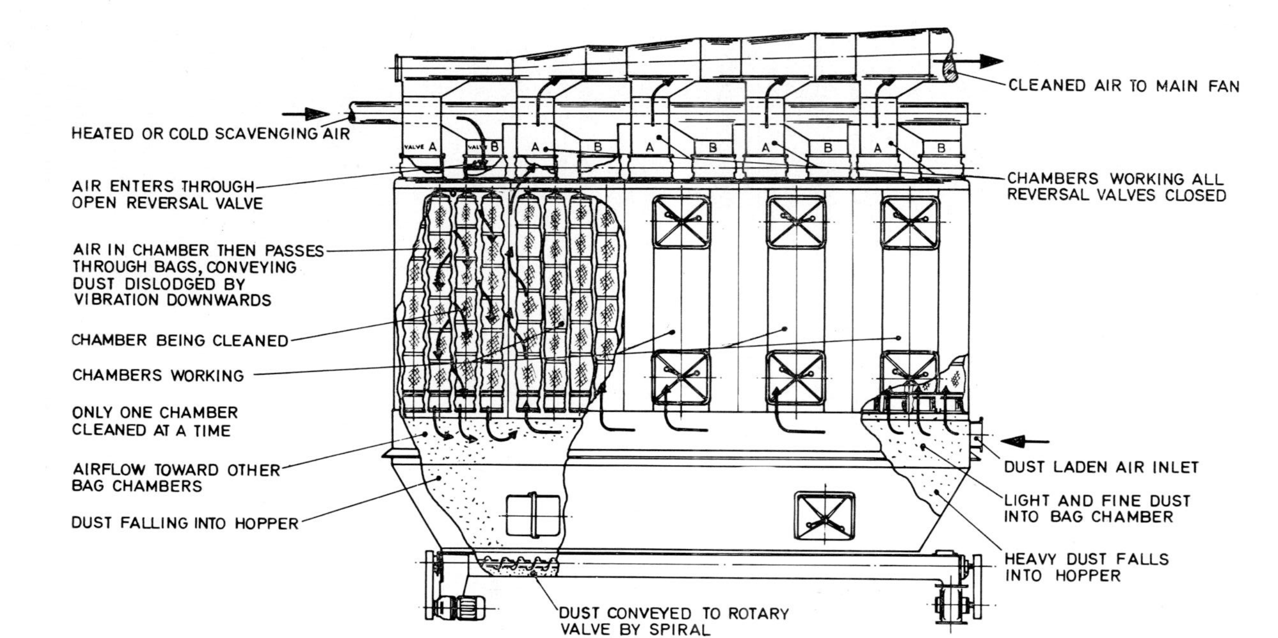

Fig 6.14 Fully-automatic fabric filter with reverse-air-flow cleaning of filter elements

available and, provided that full details of performance capabilities are available, correct application procedures should not be too difficult.

THE DESIGN OF CONVEYING DUCTS

With hood and connection design established and the total air entrainment volumes known, it is a relatively simple matter to design an orthodox and satisfactory ducting system.

Velocities throughout the system should be as uniform as possible as sudden changes waste energy and can lead to choking and abrasion. Velocities will vary between 4,000 and 6,000 f.p.m. (20 to 30 m sec^{-1}) depending on the dust and swarf to air ratio.

Correct design and specification of material thicknesses is of prime importance for ductwork for normal systems but the advice of the specialists should be sought for any abnormal applications. The following table of indicated thicknesses may be regarded as a standard code of practice.

Bulk handling systems

	Ferrous swarf	*Non-ferrous swarf*
Bends	½ in (12 mm) Mild-steel plate	¼ in (6 mm) Mild-steel plate
Straights	¼ in (6 mm) Mild-steel plate	⅛ in (3 mm) Mild-steel plate

Combined swarf and dust systems

	Ferrous swarf	*Non-ferrous swarf*
Bends	¼ in (6 mm) Mild-steel plate	⅛ in (3 mm) Mild-steel plate
Straights	⅛ in (3 mm) Mild-steel plate	14 s.w.g. (2 mm)

If the cross section of the ductwork changes abruptly then unnecessary pressure losses will occur. For example, these can include entrance losses at the pick-up boot and at bends where the throat radius must be restricted. Bends should always be of generous radius but, if site conditions preclude this, then the bends must be made from some wear-resistant material.

Individual branch pipes joining the main duct should not be introduced below the horizontal centre line; adherence to this rule will preclude any gradual choking of such pipes if they are nof used for any length of time.

Most ductwork systems operate below atmospheric pressure and thus any failure in sealing along the duct length will result in air leaks into the system, thereby reducing the entrainment of dust-laden air at the hoods or nozzles. It is necessary to seal the ducts as effectively as possible and, even then, to allow a margin of air flow to compensate for leakage.

The strength and number of the supports for the ducts or pipes requires consideration. They should be designed on the assumption that, should choking of the duct occur, they will be able to support the increased load.

TYPES OF PNEUMATIC CONVEYING SYSTEMS

The actual basic design of a pneumatic conveying plant, particularly in respect of an automatic transfer line with many machining stations, can take two forms. Firstly, the swarf and dust can be entrained pneumatically by individual hooding of the cutters in the cutting zone—see Figure 6.15. Alternatively, the swarf can fall by gravity into the base of the machine and can then be transported by means of a mechanical conveyor to the end of the transfer line where it can be deposited into the suction pick-up boot of a pneumatic conveying system. The combination of close-proximity hooding and mechanical conveyors underneath the machining stations can offer maximum swarf clearance, close-proximity hooding catering for the majority of the swarf and dust from the cutting zone, the mechanical conveyor dealing with any swarf that is thrown away from the hooding by the cutting-tool action.

Individual machine tools can also be quite readily adapted so that swarf and dust extraction are combined in one plant. In this case, there is usually space available to accommodate the combined pneumatic-conveyance and dust-extraction equipment locally at the machine tool and very successful applications have been carried out in this field—see Figure 6.16. There is no distinct line of demarcation between a dust-extraction system with moderate

Fig 6.15 An example of exhaust hoods applied to milling cutters

Fig 6.16 A slideway grinding machine equipped with a self-contained pneumatic swarf and dust collection system

weights of material conveyed per pound of air and the pneumatic conveying system operating with higher material-to-air ratios. That different types of system are distinguished is due chiefly to the fact that a centrifugal fan is normally used with low-density systems whereas a high-pressure fan or blower is required for the higher density ratios, with multi-stage air turbines being applied to the low-volume/high-velocity method of conveyance.

In order to illustrate current methods of pneumatic handling and the advantages to be gained when applied to machine tools, it is necessary to consider three principal systems, as follows:

(a) *Low-volume/high-velocity systems*
(b) *Pneumatic systems for the removal of airborne and inert swarf.*
(c) *Pneumatic removal of swarf from individual machines by unit collectors.*

Each system is considered separately as follows:

(a) Low-volume/high-velocity systems

For certain automatic-transfer-line applications where the size of component restricts the space available for hooding, a system known as 'low-volume/high-velocity' is used. The basic design utilises the minimum volume of air for swarf removal and conveyance, thus allowing the use of pipes of small diameter and these can usually be accommodated in the most complicated layout.

High-velocity air movement is achieved by high suction or 'vacuum', the velocity being created by the passage of relatively small volumes of air through small-bore pipes and nozzles. Because of this, fans that are commonly associated with other dust and swarf extraction and collecting systems are not suitable for this application.

The rate of air flow required varies with each application and depends on many factors but, for the purpose of comparison with other methods, the low-volume requirements can be as small as 10 cu ft (280 cm^3) per minute with a maximum of 100 cu ft (2.8 m^3) per minute per point. Nozzle velocities are also variable but it would be preferable to consider 6,000 to 12,000 ft (1,800 to 3,600

m) per minute as a design basis. With extraction pipes ranging from $\frac{3}{8}$ in to $1\frac{1}{2}$ in (10·5 to 38 mm) diameter, the system is particularly suitable for applying to machine tools when the saving of space is of prime importance.

When using extraction pipes having bores within the above range, turbo-exhausted units are necessary as the suction rating is usually above 30 in (750 mm) water gauge. The general level of suction requirements is seldom below 60 in (1,500 mm) water gauge and where this value is necessary in the nozzles, the additional lines of conveying ducts and separation unit can increase the over-all suction requirements to 120 in (3,000 mm) water gauge. It is preferable, however, to keep the over-all suction requirements to 100 in (2,540 mm) water gauge in the interests of economical running. With higher suction power, or 'vacuum' as it is more commonly expressed in these ranges, the inlet velocities increase and extra care must be exercised to prevent an out-of-balance characteristic occurring in a complex layout, i.e. increased air volumes at some nozzles due to actual nozzle resistance.

Air movements of this kind have previously been associated mainly with dust extraction from portable tools such as grinders and chisels but the performance range of available plant is equally applicable to swarf and every kind of cutting or abrasive action on material such as stone, asbestos, coal, graphite, precious metals, explosives and minerals. Metals and minerals which, in swarf form, can be considered friable when handled, can be extracted successfully through the small nozzles which are referred to later.

Figure 6.17 shows how swarf collected on a cast-iron workpiece being machined on a vertical boring mill and Figure 6.18 shows how the swarf is removed when a low-volume/high-velocity (LV/HV) nozzle is applied.

The position of extraction nozzles must be carefully arranged as it is important to recognise that design in this performance range is not an extension of the practice of normal dust extraction and removal systems. Consequently, 'hooding' which is fitted around the workpiece or cutting tool, in close proximity, is not a collecting chamber with an outlet port to a duct. When operating with low volumes, the work is done as a result of high-velocity air movement

Fig 6.17 Shows how swarf will collect on a cast-iron workpiece being machined on a vertical boring mill

Fig 6.18 Shows how swarf is removed when a low-volume/high-velocity nozzle is applied

and the position of the nozzles is secondary to the need for the air jet to pass over the workpiece.

The effective range of high-velocity air must be accepted as extremely short and the best results will always be obtained when the jet of air into the nozzle is taken from an enclosed space, whether in the form of a nozzle housing or a hood developed to provide the air inlet at the most strategic air-entry position. Nozzles with low-volume air displacement can be used to create air glands in which the air movement is made to flow round the workpiece and so prevent the release of swarf and dust. This method is suitable only with small workpieces and Figure 6.19 shows a typical example.

Fig 6.19 A typical example of a nozzle housing surrounding a workpiece. Neither workpiece nor cutting tool can be seen

Contrary to the foregoing, when it is not possible to apply external nozzles, aeration sleeves may be fitted to allow a low volume of air to pass through the workpiece and thus entrain the swarf and dust particles from the cutting or abrasive action. If the quantity of swarf produced is likely to be beyond the capacity of the extraction nozzle, which may be small because of space restrictions, then external aeration sleeves will provide the extra air necessary for the entrainment of the excess swarf and dust.

Swarf which cannot be arrested at source can be collected by gravity in small hoppers or trays and, with aeration sleeves, be removed through the conveying lines.

On transfer machines for drilling, threading and facing, singly or on multi-spindle machines, the extraction is effected by any of the above applications, and nozzles of various shapes and sizes may be used in order to remove swarf from inaccessible areas; aeration sleeves can be provided as channels, ducts or arteries in bush plates or bed plates. Figure 6.20 illustrates the adaptability of LV/HV application to an automatic transfer line handling small non-ferrous components.

Automatic lathes and milling machines with high outputs of swarf can be designed with the swarf-extraction ports within the machine and with externally connected vacuum service lines; the suction can be switched on and off by mechanically aided valves

Fig 6.20 Illustrating the adaptability of a low-volume/high-velocity system to an automatic transfer line for machining small, non-ferrous components

using the cutter movement as the actuator, so reducing over-all air consumption.

Traversing nozzles of simple form, actuated by tool-holders or other levers, can be shaped to cover surfaces and sweep up swarf and dust where deposits are trapped by the parent piece of material whilst in the cutting position.

To meet average requirements, it is reasonable to allow 0·5 h.p. per nozzle and for each conveying duct coupled to manifolds distributed at strategic points around a machine tool; where several machine tools may be served from a single extraction plant, the total horse-power should be increased by 50% after making an assessment of the number of inlet nozzles required.

The LV/HV basic performance is similar to that of a vacuum cleaner. Performances of 30 in (750 mm) water gauge and 30 c.f.m. (0·85 $m^3 min^{-1}$) or 100 c.f.m. (2·8 $m^3 min^{-1}$) at 100 in (2,540 mm) gauge are all within the working duties of industrial vacuum cleaning plants which combine separators, both primary and secondary, with motorised suction units. Generally of a simple form, these plants are available both as mobile or static units using motors between 0·5 h.p. and 15 h.p.—see Figure 6.21.

Fig 6.21 This is a standard industrial vacuum cleaner that is capable of operating one high-velocity extraction nozzle and having a capacity of 40 c.f.m. (1.13 $m^3 min^{-1}$) at 35 in (88.9 cm) water gauge

In the static-plant range there is a wide choice of equipment available. The largest standard vacuum producer of the multi-stage turbo type is 150 h.p. at the present time but larger capacities are due to be produced in the near future.

Vacuum producers with alternative features are available but those of the positive-displacement type are less flexible in application and demand far more care against breakdown due to dust carryover. The multi-stage air turbine is a frictionless unit requiring little maintenance. Noise levels are low enough to enable the units to operate without silencers but these should be added to the exhaust outlets where the discharge terminates in built-up surroundings.

Separation units for large-capacity installations are usually designed to specific requirements to provide storage capacity within the primary separating vessel as dictated by the demand of the system and the user's required method of swarf disposal; containers of several tons capacity are not unusual. Secondary separators, in the form of porous filters, are used to clean the conveying air, and the casing of the filter with dust-storage hoppers can be supplied with manual or automatic filter-shaking mechanisms. Atmospheric air-reversal filter-cleaning methods using the vacuum

Fig. 6.22 Twin filter separators where the filter cleaning is done by atmospheric air reversal, using the vacuum within the system

existing within the system are sometimes preferred but in these cases twin filter compartments are essential if the air-reversal action is to be effective—see Figure 6.22. Such separators are mostly of cylindrical form and must, of necessity, be designed and constructed in a way that will enable them to withstand the load created by the maximum possible vacuum or negative pressure developed by the vacuum producer. As these separators are in the nature of pressure vessels they will need to be examined regularly for insurance purposes and records of inspections maintained for the Factory Inspector. Where discharge valves are used in conjunction with separators, they should be sufficiently strong to offer protection against leakage under the load and be able to carry the weight of material and resist the abrasive action.

(b) Pneumatic systems for the removal of airborne and inert swarf

A composite pneumatic conveying system involving close-proximity hooding of cutting-tool points may be used with a mechanical conveyor fitted underneath the transfer line. Swarf which is not picked up by close-proximity hooding surrounding the cutting tools is collected by the conveyor and deposited into a suction pick-up boot connected to the main pneumatic system. In this way most of the swarf may be transported to a centralised disposal zone. Such a system is very well suited to meet the requirements of a machine tool producing cast-iron swarf in large quantities.

The removing and transporting of swarf under these conditions is a problem that is usually associated with transfer lines or lines of machines using the same material. The problem may be overcome by using any of the following systems:

1 an arrangement which relies solely on the efficiency of the close-proximity hoods to collect all the swarf as it is produced in the cutting zones, and an adequate air volume and velocity to convey it through suitable ducts to the disposal area;

2 a system which relies on gravity and manual removal of the swarf from the cutting zones to a mechanical conveyor which carries it to a suction pick-up boot from whence it is pneumatically conveyed to the disposal area;

Fig 6.23 Individual fabric unit applied to a milling machine. Note the close-proximity hoods around the cutter, the additional vacuum clean-up point, and the acoustic cover over the fan on top of the collector

3 a combination which uses close-proximity hooding and air movement to collect and transport the swarf from the point at which it is created, i.e. the cutting tool, together with a mechanical conveyor sited so that it will receive the swarf that may escape the close-proximity hoods and which falls by gravity to the base of the machine. Such a conveyor will be used to transport the swarf to a suction pick-up boot where it may be entrained in the airstream for conveyance to the disposal area.

Each application must be treated on its own merits, taking account of the following basic considerations:

1 size, shape of components to be machined;

2 size, shape of fixture to be utilised;

3 diameter, speed and type of cutter to be used;

4 particle size and shape of chip produced and rate of chip production per hour;

5 type of material to be handled, density of same;

6 direction of motion of chip from cutting point;

7 the possibility of liquid cutting fluid being used.

When the foregoing information has been obtained and recorded, the designer should be in a position to prepare an outline scheme for discussing with all concerned, including the personnel responsible for operating the system. It is at this stage that the success or failure of a scheme may be decided; too much stress cannot be laid on the need for joint consultation between all parties so that unanimous approval can be gained.

It is not possible to give accurate capital cost figures for bulk-handling pneumatic conveying systems because of the wide variation in equipment, duct specification and actual length of conveying duct.

(c) Pneumatic removal of dust and swarf by individual unit collectors

For swarf removal from machine tools where a combined pneumatic conveying and dust-extraction application is essential, close-proximity hooding is again the optimum solution, with the collection equipment being an individual swarf and dust collecting unit situated close to the machine tool.

Users of machine tools are often faced with the problem of applying swarf and dust control and removal facilities to individual machines which cannot be served by existing plant installations. In such instances the use of unit collectors offers many advantages and, in certain circumstances, they may be used throughout a machine shop as an economical method of collecting swarf. This type of collector is of relatively limited capacity although, if it is located near to the source of the swarf and dust, it may serve two machine tools; however in the case of bigger machine tools that

Fig 6.24 Individual fabric units in use on a line of cylinder boring machines

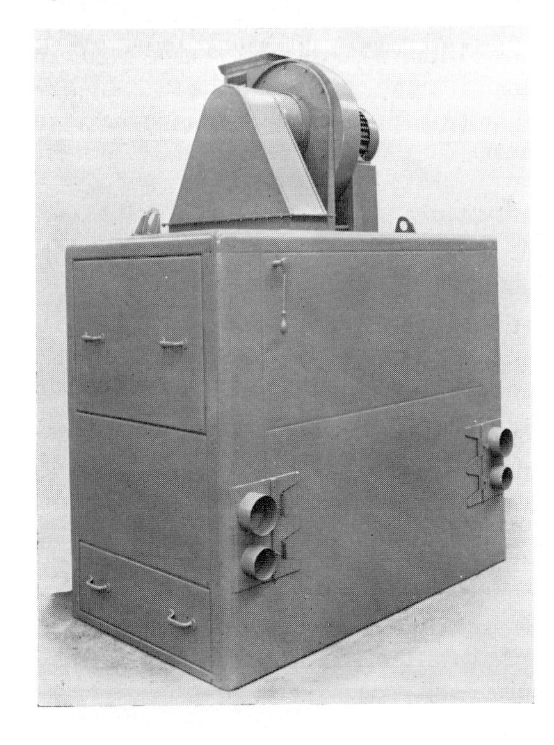

Fig 6.25 A fabric-type filter unit which embodies an automatic shaking cycle for cleaning the fabric filter elements

produce large volumes of swarf it may be necessary to use more than one unit collector.

A system employing a unit collector normally comprises three basic sections:

1 hoods—these are common to all types of installation and have been described in some detail;

2 inter-connecting ductwork, because of the close location of the unit collector to the machine tool, is of limited extent and presents minimum resistance to the flow of air;

3 the unit collector itself which may be designed as a wet de-duster or a dry fabric filter, the latter being more usually used in this type of installation.

A number of well-designed unit fabric filters is available; most of them are provided with a filter, fan, hopper and swarf and dust receptacle, self-contained within a chassis; and a range of filter and fan capacities is offered. For an application to be successful, the following parameters must be established and related to the capacities of the unit collector:

1 the air-entrainment rate required;

2 the characteristics of the dust and swarf to be entrained, i.e. quantity, size distribution, and particle shape;

3 the period of continuous operation is required;

4 the ultimate dust and swarf disposal facilities required.

A detailed analysis of the foregoing factors is a matter for the specialist but the following principles generally apply:

1 The air-flow capacity of the unit will be determined by the filter area and the fan characteristic. This should be calculated in consideration of the swarf load to be handled.

2 Whilst most unit collectors employing fabric filtration have automatic cleaning devices, they are basically intermittent in operation, i.e. they must be stopped periodically to allow for clean-

ing of the filter element under conditions where no air is flowing. Under existing conditions, it is likely that opportunities for cleaning would occur four times during an eight-hour working day; consequently, the filter unit collectors should be selected on the

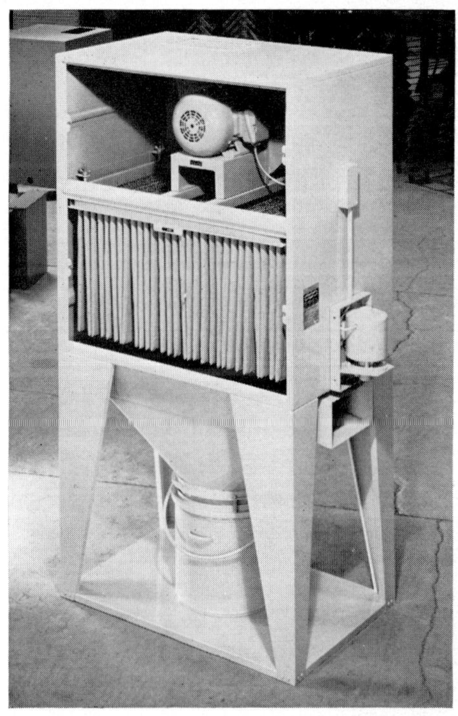

Fig 6.26 Another type of fabric-filter unit which embodies an automatic shaking cycle for cleaning the fabric filter elements

basis that they will operate effectively over a two-hour period before cleaning is required.

3 In view of the conditions outlined in 1, it follows that the application of fabric-type unit collectors will be most satisfactory when the dust and swarf are in relatively low concentrations.

4 If continuous operation is required then a continuously rated filter is essential; another method of obtaining continuous operation would necessitate duplicating the unit collectors.

5 The storage and disposal of the collected swarf and dust must be arranged so that operation of the plant is not affected. Because

the quantities of swarf and dust are usually small, designs which include bins, drawers or sacks for receiving the swarf and which can be emptied when the installation is not in use, normally have adequate holding capacity. There are more sophisticated systems for removing the swarf and dust from the unit collectors but these are mainly associated with continuously rated operation and involve ancillary equipment such as continuous rotary discharge valves, etc.

Fabric-filter unit collectors, correctly applied and maintained, offer a high degree of collection efficiency and, when non-toxic materials are involved, it is possible for the filtered air to be released within the workshops. It is advisable, however, to consult H.M. Inspector of Factories before this is put into operation.

Although wet unit collectors offer a high degree of separation they are not so efficient as fabric-type collectors. Because of this and the increased humidity, the air should not be re-circulated in the workshops. The collected swarf sludge may be removed automatically by drag-link-type conveyors or manually during the periods when the plant is not operating.

SUMMARY

A well-designed and efficient pneumatic conveying system offers many benefits and some of these are as follows:

1 It has the ability to convey swarf and dust in any direction and over any reasonable distance.

2 It affords total enclosure of the swarf whilst it is being transported.

3 Close-proximity hooding can remove swarf from the cutting zone just as rapidly as it is produced.

4 The additional advantages of in-built dust control make for better housekeeping.

5 The number of moving parts is generally limited to the air actuator and the separation equipment and so maintenance is restricted to these items.

6 There is a saving of floor space as usually the collection unit is sited away from the production area and invariably the ducts rise vertically from the machines to an overhead main.

7 If exhausting is to take place beneath the machines then only a limited amount of ground preparation is needed.

8 The removal of the swarf and dust as soon as it is produced means less maintenance work on the machine slideways, bearings, etc., so increasing the life of these parts.

9 Having in mind the health of machine operators, pneumatic removal systems are the only ones that make an all-round improvement in working conditions.

As with all systems, pneumatic conveying has certain disadvantages that are relatively small, these being as follows:

1 A pneumatic conveying plant can deal with wet swarf only when the moisture content does not exceed 6%. Centrifuging of swarf prior to pneumatic conveyance is essential when the moisture content is above this limit. Close-proximity hooding is practical only where the moisture content does not exceed 6%.

2 It cannot deal with bushy, curly swarf unless previously crushed to a suitable size and shape.

3 Power costs are higher in comparison with mechanical conveying of similar capacity.

4 The noise factor—some noise is inevitable and special precautions may be necessary to reduce it to an acceptable level. The pneumatic-conveying engineer is constantly developing new techniques to keep noise down to a minimum and, in conjunction with acoustic engineer, can put forward suitable silencers and acoustic ducts to reduce the fan outlet noise-level problems to an acceptable limit.

5 Individual collector applications require frequent removal of collected swarf and hence use of manpower.

The information given in this chapter represents a serious attempt to enlighten the machine-tool designer, the machine-tool user and all other parties interested in the efficient removal of swarf from all kinds of machine tools. It must be appreciated, however, that only a broad outline of the possibilities of pneumatic conveying can be given within the confines of this manual. By reason of its somewhat specialised nature—it is sometimes regarded as more of an art than a science—it is essential that the advice of experts having many years of practical experience be sought when consideration is being given to the introduction of pneumatic conveying. This will enable a complete evaluation to be made of the circumstances and difficulties and should ensure that all problems will have been recognised and overcome before the system is installed.

The scope for pneumatic removal and conveying of swarf in the engineering industry is vast if only the design engineers of the machine tool and pneumatic engineering industries will work together towards a common end. Adequate swarf removal by pneumatic means can be made an integral part of almost any machine-tool design where dry cutting prevails and bushy swarf is not produced.

7 CUTTING FLUIDS AND SWARF

INTRODUCTION

With wet machining, as distinct from dry machining in which the presence of cutting fluid is not required, the problem of handling swarf becomes more complex. Large chips tend to retain a small proportion of cutting fluid in the form of small globules on the surface of each chip, and small fragments of swarf, including dust, tend to settle in or on the cutting fluid, causing sludge or a metallic scum. In the case of single machine tools the swarf is normally raked away from the machine, and swarf plus cutting fluid falls on to the shop floor or is removed to a storage zone in which swarf and cutting fluid tend to separate. The cutting fluid soaks into the ground or through an appropriate outlet and the swarf, usually ferrous, begins to rust.

On many machine tools of the automatic-transfer type, cutting fluid is used not only as an aid to machining but also as a means of washing swarf away from the cutting zone into a duct in which the swarf is transported hydraulically to a convenient take-off point. Such a system is usually one which requires a large volume of cutting fluid to handle a continuous through-put of swarf and the problem of separating swarf from cutting fluid, before recirculating the cutting fluid is, therefore, a real one. Usually, the problem is tackled in two stages. Firstly, large swarf particles are removed at the take-off point by means of some form of scraper-conveyor which deposits the swarf into leak-proof hoppers making it easy to transport to centrifuges in which swarf and cutting fluid are separated. Secondly, cutting fluid remaining in the system is passed through a filter or clarifier which removes most of the smaller swarf particles before the cutting fluid is pumped back to each of the cutting stations.

Clearly, there are many different requirements presented by a wide range of metal-cutting situations which make use of cutting fluid and it is not easy to specify how to deal with the individual requirements of each situation. But the properties of cutting fluids are clearly of direct interest in connection with the choice, design and operation of swarf-removal equipment and in this chapter a brief account is given of these properties and of the factors that have to be considered in the design of swarf-removal equipment.

THE FUNCTIONS OF CUTTING FLUIDS

In order to meet the demands of increased production at lower costs, machining conditions are increasing in severity with the consequence that maximum tool life is of prime importance. Improvements are continually being made in machine tool and cutting-tool materials in order to cope with these conditions, but to achieve optimum performance it is essential to give careful attention to the selection of the cutting fluid to be used. For the correct selection of cutting fluids it is essential to have an understanding of the way in which a cutting fluid functions and a brief account of these actions is given below.

It is generally accepted that the primary functions of a cutting fluid are to cool and lubricate the tool, and thereby:

1 To increase tool life or permit higher metal-removal rates.

2 To produce a good surface finish.

3 To reduce power consumption.

In addition to these primary functions a cutting fluid performs a number of important secondary functions such as:

4 To cool the chip and workpiece and hence to assist in maintaining the accuracy of the finished workpiece.

5 To carry swarf away from the cutting zone.

In designing a cutting fluid to perform these various functions care is taken to ensure that no undesirable side effects are encountered and in particular that the cutting fluid is:

(*a*) non-corrosive to workpiece and machine-tool components;
(*b*) stable during its service life;
(*c*) non-toxic and non-irritant.

For certain specialised applications cutting fluids may be required to have other particular characteristics such as transparency and compatibility with machine-tool lubricants.

The influence of cooling on tool life will be readily understood. Good cooling requires a fluid of high specific heat, so that water is an obvious constituent where it can be used. However, although oils have only half the specific heat of water, they have cooling properties that are adequate for many jobs, especially when of moderate viscosity and properly directed in ample quantity.

The most effective form of lubrication is for the two sliding surfaces to be completely separated by an oil film. However, due to the high pressures and resultant high temperatures involved in the majority of cutting operations this condition is seldom encountered and therefore load-carrying and friction-reducing additives have to be incorporated. Although the precise mechanism by which cutting fluids penetrate to the regions of highest pressure is not fully understsood, practical tests and field experience have proved that additives incorporated into cutting fluids are effective in prolonging tool life and maintaining a good surface finish.

The load-carrying additives used in cutting fluids fall into two main classes—boundary and extreme pressure. Boundary lubrication depends on films which are only a few molecules thick and cannot completely separate the surfaces at all points. Under these conditions polar fatty additives are used which adsorb on to the sliding metal surfaces and lower the coefficient of friction. With more arduous machining operations a point is reached where the boundary lubricating film breaks down, and gross metal-to-metal contact occurs. Under these conditions welding is prevented by the use of extreme-pressure additives. The extreme-pressure additives, usually based on sulphur or chlorine, react with the metal surfaces to form low-shear-strength compounds between the chip and tool. These low-shear-strength compounds are generally accepted to be effective up to the temperatures at which they melt; in the case of chloride films this temperature is in the region of 400°C and in the case of sulphide films is in the region of 800°C. Chloride films generally give lower values of friction than sulphide films but, because of their lower melting points, chloride films may not be so effective as sulphide films in high-speed machining operations. In some cases beneficial results are obtained by using a combination of chlorine and sulphur-based additives. Extreme-pressure additives based on compounds other than chlorine and sulphur, such as, for example, phosphorus, are occasionally used.

TYPES OF CUTTING FLUIDS

The diversity of cutting operations, materials and types of work-piece, means that a very wide range of cutting fluids is necessary to ensure that the best machining conditions are always obtained. The main types of cutting fluids are summarised briefly in the following paragraphs and their chief characteristics and uses are indicated.

1 *Soluble cutting fluids*

There are many metal-cutting applications where the principal property required of the cutting fluid is high cooling ability. Water, because of its high specific heat and latent heat of vaporisation, is an excellent cutting fluid but has poor lubrication and anti-corrosion properties. Soluble cutting fluids combine the high cooling ability of water with the good anti-corrosion properties of mineral oils and in addition possess better lubricating properties than water.

Soluble cutting fluids fall into two main groups, namely soluble oils and chemical solutions.

(*a*) *Soluble oils* Soluble oils generally consist of blends of mineral oil, emulsifying agents and other additives which stabilise the emulsion and prevent corrosion when the oil is mixed with water. Soluble oils is a misnomer since when mixed they consist of a suspension of oil droplets in water. The oil-droplet size determines whether the emulsion is of the milky or translucent type.

Emulsions provide maximum cooling with limited lubricating properties as they are used at high dilutions, typical water to oil ratios varying from 10:1 to 30:1 for general machining and 60:1 to 80:1 for grinding operations.

To enhance their lubricating properties soluble oils may also incorporate fatty oils or extreme-pressure additives.

(*b*) *Chemical solutions* Chemical solutions consist of mixtures of water-soluble corrosion inhibitors and surface-active load-carrying materials. These fluids may be intended for use at high dilutions only as grinding fluids or, in the case of the more sophisticated formulations, as general-purpose cutting fluids for both machining and grinding according to the dilution used.

2 Neat cutting oils

These are used for operations which demand greater lubricating properties than can be provided with a water-mixed cutting fluid or where the design of the machine tool cannot prevent contamination of the lubricating and/or hydraulic system with cutting fluid.

Neat cutting oils fall into several main categories as listed below.

(*a*) *Straight mineral oils* These range in viscosity over the BS 4231 grades 7 to 32. They are generally suitable for light machining operations on free-machining steels, yellow metals and aluminium alloys.

(*b*) *Compounded mineral oils* These are blends of mineral and fatty oils. The fatty-oil constituent provides improved lubrication under the boundary lubrication condition referred to earlier. They are used for similar operations to those listed for straight mineral oils but have improved wetting properties and would be expected to give improved surface finish and increased tool life.

(*c*) *Extreme-pressure cutting oils* These are blends of mineral oil and extreme-pressure (EP) additives in varying combinations and quantities. The most common types of EP additives in use are chlorinated waxes and paraffins, sulphurised fatty oils and elemental sulphur.

Those blends which contain elemental sulphur are generally referred to as 'active' extreme-pressure cutting oils as they will stain cuprous metals at room temperatures, blends of extreme-pressure additives other than elemental sulphur being known as 'mild' or 'inactive'.

The choice of the viscosity of the base oil, combination of additive types and concentration determines the level of performance. Most cutting-oil suppliers provide a range of neat cutting-oil blends to cater for the full range of machining operations and materials experienced in industry.

(*d*) *Special-purpose oils* For certain specialised applications it is necessary to provide neat cutting oils having properties not necessarily required of neat cutting oils for general-purpose machining.

Examples of this are deep-hole boring and broaching. For deep-hole boring a low-viscosity oil of high additive content is required because of the high metal-removal rates and the need for the cutting fluid to carry the swarf for comparatively long distances through confined spaces. For broaching, low-viscosities are essential because of the need for the cutting fluid to flow readily to the cutting zone and, in addition, a high level of reactivity is required because of the low temperatures realised in the cutting region in the broaching operation.

Metallurgical considerations may also require the use of particular types of cutting fluid in order to avoid stress or temperature corrosion problems.

SWARF, CUTTING FLUIDS AND HANDLING AND PROCESSING EQUIPMENT

Emphasis has already been laid on the primary functions of cutting fluids and attention has been drawn to the assistance that can be given by cutting fluids in the clearing of swarf from the cutting zone and flushing it from the machine surfaces, but the swarf has still to be removed from the actual machine tool. In previous chapters, much of the equipment normally used in conjunction with swarf handling has been described and illustrated in detail but the following information is very relevant and should be borne in mind.

Materials of construction

The materials of construction of those parts of machine tools that are likely to be in contact with the cutting fluid, including the pump, must be selected to resist attack by all types of fluid. In particular, the following materials should be avoided:

Yellow metals, light alloys, zinc alloys and natural rubber. If these recommendations are considered to be too sweeping they can be sub-divided as follows:

Where neat cutting oils are used, avoid yellow metals and natural rubber.

Where soluble-oil emulsions are used, avoid light alloys, zinc alloys and natural rubber.

These recommendations apply particularly to cutting-fluid pumps and feed pipes, but could, with advantage, be extended to all parts of machine tools likely to come into intimate contact with the cutting fluid.

Finishes

In their development work, cutting-fluid manufacturers pay great attention to ensuring that their products do not attack properly applied machine-tool paints. Complaints from machine-tool users of paint removal do occur, however, and it is recommended that machine-tool manufacturers should ensure that, for all painted surfaces, the correct paint is used and that it is correctly applied and cured.

TREATMENT OF CUTTING FLUIDS

There is an increasing awareness of the benefits to be gained and the economies to be made by using cutting fluids that have been subjected to a filtration process that removes a large proportion of swarf particles and other contaminants. In this connection there are certain factors that merit attention:

1 When selecting filtration equipment, the user should consult the manufacturer, being very precise in specifying his requirements and stating the nature of the cutting fluids the equipment must process, i.e. type, viscosity, volume to be used, etc. This is particularly important with neat cutting oils where viscosity has a significant effect on filterability, especially so with magnetic filters and gravity separators.

2 In selecting equipment, consideration should be given to the general design. Exactly as with other cutting-fluid equipment, filtration and separation appliances should be easy to clean and maintain.

3 When the selected appliance incorporates a filter medium, the advice of the manufacturer should be sought so that the most suitable medium is used. This is important as some chemical media will remove active ingredients from the cutting fluid.

4 Centrifuges are effective instruments for the separation of swarf particles from cutting fluids but it should be noted that not all centrifuges are suitable for dealing with all types of cutting fluids. It is again recommended that the requirements be fully discussed with the manufacturer before a decision is finally made.

5 Hydrocyclones are becoming increasingly popular for incorporating into cutting-fluid systems in grinding machines. Selection is very important as the effectiveness of a hydrocyclone depends on the viscosity of the cutting fluid used.

6 Because cutting fluids absorb some of the heat generated during the cutting operation and in the working parts of the machine tool, it is sometimes decided to incorporate some form of refrigeration in the system. This must be done with caution as trouble may occur because of the removal of active ingredients from the cutting fluid through the reduction of additive solubility which occurs at low temperatures.

7 In large central systems, the oil must be periodically examined

to ensure that it is fit for further use. Oil that has been continually recovered may lose the ability to perform the duties for which it was devised, although normal make-up with new oil should compensate for this.

8 In recovery systems that include apparatus for pasteurising the cutting fluid, care must be taken that the heating units are suitably protected by sheathing made from a non-corrodible material. The surface rating should be selected with care as local overheating can cause decomposition of the oil.

9 The grinding of grey and nodular cast iron by a wet process can cause difficulties with equipment. Amongst the swarf produced by grinding will be minute graphitic fines that, when recirculated, tend to bring about an early breakdown of the cutting fluid. Also, the finer, powdery swarf particles suspend themselves in the cutting fluid, even to the point of almost dissolving, and will quickly plug filter media, rendering them useless.

10 The condition of the water used in preparing an emulsion could be important and, particularly when a central system is to be used, should be analysed. Soft water, having a hardness of less than 100 parts per 1,000,000, will cause foaming and this is something the cutting-fluid manufacturer tries hard to avoid. On the other hand, hard water, having a hardness of more than 250 parts per 1,000,000, tends to form insoluble soaps that will quickly clog filters, load grinding wheels, leave deposits on machines and tooling, etc.

11 Cutting fluids are a combination of ingredients brought together with a definite function to carry out. It is wise, therefore, not to interfere with their composition by putting in more additives.

SWARF, CUTTING FLUIDS AND HEALTH AND SAFETY

The use of large volumes of cutting fluid has been advocated as a useful means of removing swarf from the cutting zone on a machine tool and even as a suitable medium for transporting swarf to a disposal point but care and attention must be exercised to protect the personnel using the cutting fluid. The need to maintain all the cutting-fluid equipment in a clean and serviceable condition cannot be over-stressed and users must be made aware of the dangers that can be encountered from careless and indifferent handling.

Microbial growth brings about chemical changes that affect the lubricating properties and viscosity of water-based cutting fluids and may also nullify the effects of additives that are included to aid cutting, increase film strength and inhibit corrosion. High microbial activity can be responsible for poor surface finish, decreased tool life, fouled grinding wheels, odours and discoloration, and rapid corrosion after machining is closely associated with bacterial growth. Recent investigations have shown that an oil-emulsion system can contain over one ton of bacteria to every 20,000 gallons (91,000 litres) of liquid and that this content may consist of many hundreds of different species, the usual breeding source being slimes, gums, deposits and traces from previous emulsions, waste-food deposits, etc. It is characteristic of bacterial problems that they constantly worsen and useful cutting-fluid life becomes progressively shorter as a flora of microbes becomes established.

With oil emulsions, the answer to the problem lies mainly with the equipment provided. Implementation of the following suggestions will greatly assist:

1 The prompt removal of swarf from the cutting fluid is essential. If this is not done, then the swarf helps to deoxygenate the cutting fluid and this encourages anaerobic sulphate-reducing organisms to develop.

2 Tanks should be readily accessible to facilitate cleaning and this should be thorough and regularly carried out.

3 It is important to ensure that there are no parts of the cutting-fluid tank and system where cutting fluid may stagnate as such stagnation will encourage bacterial growth.

If neat oils are used as the cutting fluid in place of emulsions, the situation is not so difficult as they do not offer the same problems with regard to bacterial degradation. However, should the cutting

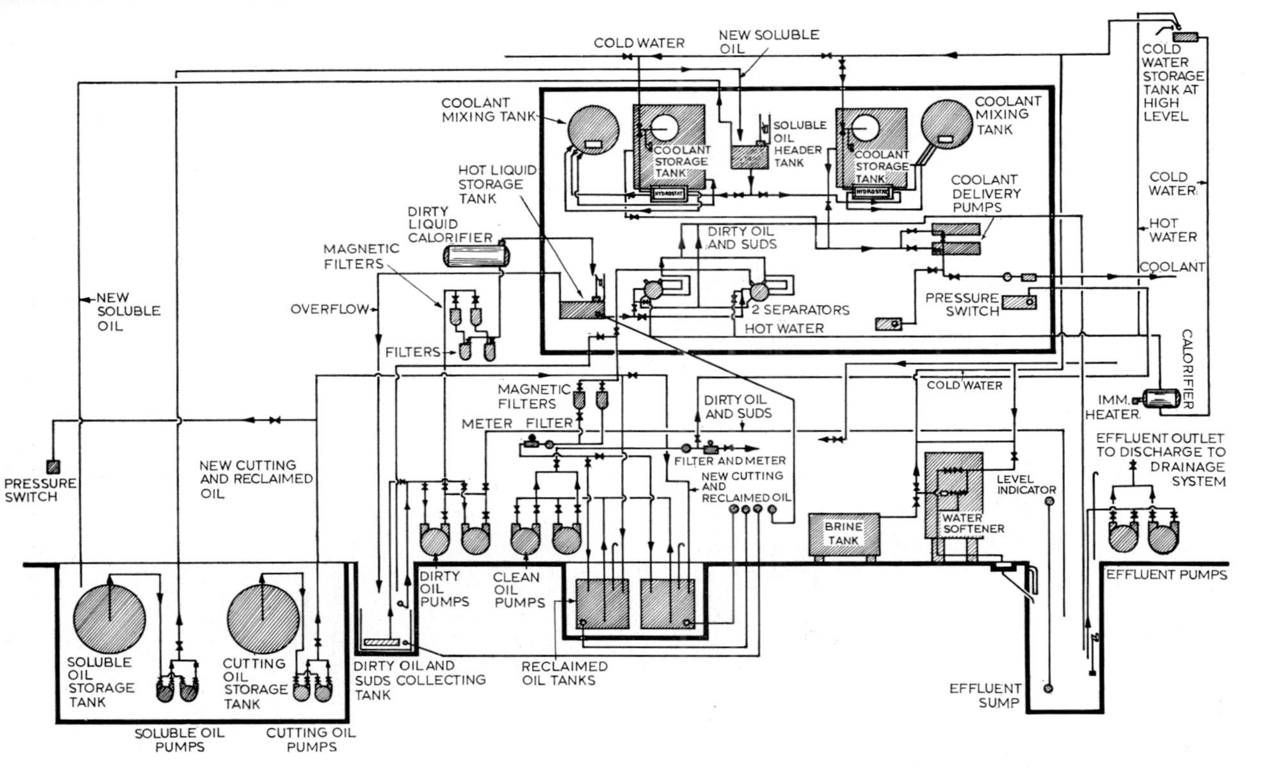

Fig 7.1 A schematic arrangement of a reclamation and central supply system for cutting fluids

oil contain any free water or become associated with any aqueous phase, then it will be necessary to take the same precautions as for an emulsion.

Machine-tool users should note the following precautions to be taken in the use of cutting fluids:

1 Cutting-fluid systems should be protected so that they cannot be contaminated by waste material such as scraps of food, cigarette ends, tea leaves, etc., as this encourages the growth of bacteria.

2 When the cutting fluid is a neat oil it should be delivered in a copious supply to the cutting zone and the cutting tools should be ground to produce discontinuous chips, thus reducing fuming and lessening the fire hazard.

3 After sump cleaning, if the cutting fluid is to be returned to the machine tools for re-use, it is strongly recommended that it undergo an effective reclamation process to remove contaminants. Figure 7.1 illustrates a schematic arrangement of a reclamation system and central supply which would be very suitable for a medium to large manufacturing establishment.

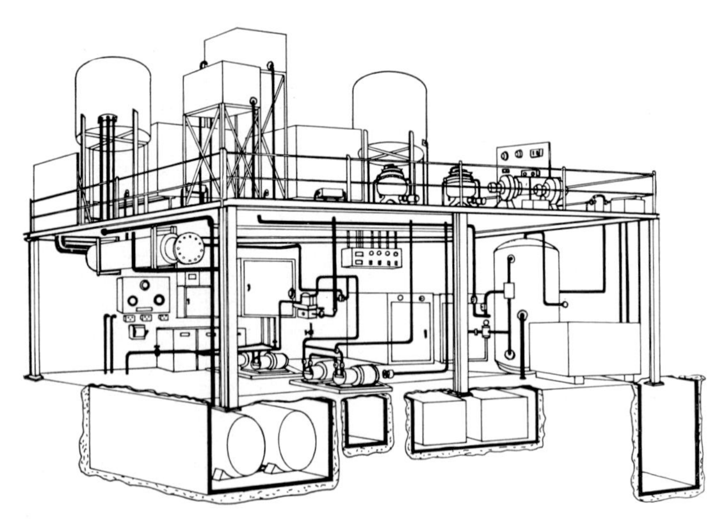

Fig 7.2 A central system for cutting fluids

4 Cutting fluids that have been reclaimed should be submitted to a pasteurising process before being re-used.

5 Cutting fluid that is not retained within the periphery of the machine will cause dirty and greasy floors; such floors are dangerous and must be avoided.

SKIN CANCER IN MACHINE SHOPS

Introduction

The knowledge that tar workers suffered from skin cancer led to the discovery in 1915 that the application of tar to a rabbit's ear could produce a skin cancer. In the 1920's, cancer, particularly of the scrotum, among mule spinners in the cotton industry was investigated; in the warm humid atmosphere of the spinning room, mule spinners wear very little clothing and oil thrown from the spindles easily reaches the skin. Research showed that solvent extracts of oils that had been heated to the temperatures of the conditions of use were cancer-forming (carcinogenic). Then investigations in the Birmingham area in the early 1950's into the incidence of cancer of the skin among workers employed on automatic bar machines led to the conclusion that some of the cutting oils used in the metalworking industry could give rise to keratoses (growths sometimes resembling warts) on the backs of the forearms, some of which became cancerous, and that there was therefore a strong association between exposure to certain cutting oils and cancer of the skin, notably cancer of the scrotum. Subsequent investigations have confirmed the connection and drawn attention to the need for action of various kinds to minimise the risk.

The Birmingham investigations

Information about the occupation of the person affected was available in about one-half of the cases of scrotal cancer recorded in the Birmingham area between 1950 and 1966. Analysis of the figures showed that there were 65 toolsetters and machine operators, 7 gas workers, 5 garage hands and 3 chimney sweeps. Other analyses show similar emphasis on toolsetters and machine operators and, in general, there would seem to be a clear-cut association between cancer of the skin and occupations in automatic shops.

The investigations also revealed that the disease can have a long incubation period. None of those affected had less than 6 years' exposure to oil and only two had less than 10 years' exposure. As bar automatics came into general use in the 1920's and the disease is occurring in men in their late 50's it is suggested that it can incubate for as long as 30–40 years after a certain period of contact has been completed, and, indeed, after all contact has ceased.

The evidence suggests that contact with cutting oil is the most likely cause of scrotal cancer in machine-tool operators—for every case of scrotal cancer attributable to soluble oil there are about four cases attributable to neat oil, although for every worker exposed to neat oil there are 4 or 5 exposed to soluble oil. But it does not follow that because there is an association between contact with neat cutting oils and cancer of the scrotum that the disease is to be found wherever neat oils are much used. In the Birmingham area, where the incidence of scrotal cancer has been shown to be high, there are many large factories in which automatics, and hence neat cutting oils, will be in use, and in which, for no apparent reason, no cases of the disease have occurred at all. Moreover, cases of occupational cancer of the skin, whatever the area attacked, form only a small proportion of the total figures for skin cancer, and it has been adduced from the evidence available that some cases of scrotal cancer will have been due to causes unconnected with oil.

Cancer of the skin

Many growths on the skin are unrelated to contact with oil. Warts on the hands are often seen, and rodent ulcers—small tumours that slowly enlarge—are found in the general population to an extent unrelated to exposure to oil. Some individuals seem to have a physiological propensity for producing tumours, and the fair-skinned, in particular, are known to develop keratoses in certain circumstances sustained over a period of time. Nonetheless, the appearance of lumps or ulcers on the skin of workers in contact with cutting oil must give cause for concern. For it is the occurrence of epitheliomata (malignant growths affecting skin tissue) at the site of keratoses in certain workers that has led researchers to the knowledge that the occurrence of keratoses forewarns that the workers are at risk.

The hands and arms of machine operators are liable to be continuously exposed to cutting oils, and other parts of the body may be contaminated as a result of splashes from or contact with the machine. There may also be exposure to oil fumes and oil mist. The adjustment, setting-up and general operation of machines and the removal of swarf are activities which bring people into contact with oil.

Recent publicity has highlighted cancer of the skin as it affects the scrotum and, in persons exposed to oil, cancer of the scrotum is more common than cancer of other parts of the skin. The evidence shows that scrotal cancer affects people in the metalworking industry whose duties involve activities similar to those of workers in mule spinning, i.e. bending over machines with friction of the legs and lower abdomen against parts of the machine such that the clothes are contaminated with oil in the area of the groin. A further source of contact arises from the use of wipers which have become soaked in oil and which soak the pockets where they are kept. The evidence also indicates that workers engaged in the collection of swarf are at risk. Scrotal cancer is not necessarily more dangerous than cancer of the skin elsewhere on the body; but the mortality rate in those who have contracted scrotal cancer is high—more because modesty tends to prevent the individual from seeking medical advice than for any other reason—and it is a more serious disease by virtue of its seat of origin than cancer of the skin on what might be termed less-vital areas. Like other skin cancers, it can be cured if treated in time.

Carcinogens

Soot, shale oil and synthetic tars have long been known to have carcinogenic properties, and powerful carcinogenic compounds—polycyclic aromatic hydrocarbons—have been isolated from coal tar, shale oil and pitch. When investigations in the Birmingham area in the 1950's led to research which showed that cutting oils were carcinogenic to laboratory animals, work was undertaken in various research centres to try to identify carcinogens in mineral oil and to discover in which fractions of oil the carcinogens might be concentrated. A report[1] published in early 1968 describes a full-scale investigation into the physical, chemical and biological properties of mineral oils. No new carcinogens were revealed but several substances were isolated that were closely related to the polycyclic aromatics—the known carcinogens—and experimental proof was obtained that carcinogens of this type might be present in uncracked oil. Tests suggested that the presence of weak carcinogens in some fractions might enhance the activity of standard carcinogens.

Strong confirmation was obtained in the tests that carcinogens are concentrated in the aromatic fraction of the oil which can be extracted with solvents; it has been established that the boiling range of the main carcinogenic fractions lies between 350°C and 450°C.

The percentage of carcinogens in the crude oil from which a cutting oil has been obtained will vary with the origin of the crude oil and the degree of refining and, as carcinogenicity is determined by experiments on animals, the identification and rating of the carcinogens is lengthy and difficult. However, there is enough evidence available for it to be said that cutting oils with a conventionally refined base will have had the carcinogens partially removed and cutting oils with a solvent-refined base will have had the carcinogens largely removed.

M.T.I.R.A. has consulted a number of the better-known oil companies and ascertained that the cutting oils they now supply—whether neat or soluble—are solvent-refined. However, some have supplied conventionally refined oils until quite recently. Inevitably the reduced-risk oils cost more to produce than the conventionally refined oils they replace.

No reliable and simple method of determining the presence or absence of carcinogens in oil has yet been developed, but the Institute of Petroleum's Advisory Committee has asked the major oil companies to try to develop, possibly as a joint effort, a method of testing oil for cancer agents.

Cutting oils

Despite the fact that cutting oils are among the commonest causes of industrial dermatitis, the majority of persons working with them do not suffer from skin troubles, mainly because the oils they contact have few irritant properties. The evidence suggests that it is not so much the oil itself as one or other of the substances added to it that gives rise to irritation of the skin. If an outbreak of dermatitis occurs it is probably due to a chlorine or other additive introduced to make the oil suitable for its particular application.

It is stated that the effect of oil on the skin varies, in general, with the boiling point of the fractions: the lower the boiling point, the more the oil is a direct irritant and capable of removing the protective fatty layer of the skin; the higher the boiling point, the greater the tendency to cause inflammation and blockage of the pores and the formation of innocent and malignant tumours.

It is known that soluble oil is safer than neat oil as a cutting fluid, and that solvent-refined oil is safer than conventionally refined oil. But not all soluble oils currently in use are solvent-refined, and it is probable that undiluted soluble oils are no better than neat mineral oils in relation to cancer hazard although an operator's clothing will absorb less oil in a given period in a situation where soluble oil is used than in another where a neat coolant is used.

A further aspect of the problem is contamination of the oil whilst in use and it is not known whether levels of safety are maintained during use and re-use. The cracking effect on the oil of the heat produced at the cutting edge of tools, and metal particles and anti-rust coatings from the components machined can contaminate the coolant and further contamination can arise from food refuse, cigarette butts, spit and even urine. The study of the bacterial contamination of oil—which is rapidly becoming an important science—has shown that bacteria can multiply so rapidly that just a few can produce gross contamination in an oil overnight, with marked chemical changes.[2] It is possible that the use of non-carcinogenic oils may induce a false sense of security if the oil is not regularly monitored and changed.

Prevention

Whether it is clean or contaminated oil that is to blame steps can be taken to minimise the risk to workers and some of those applicable to machine shops are set out below.

Reduction of exposure Some of the factors which predispose to oil-contaminated clothing in machine setters and operators and in swarf collectors have been under investigation for a long time, and progressive developments in technology have produced changes in the nature of some of the jobs of those at risk that augur well for the future. The elimination of pockets where swarf and coolant can lodge or collect on machine tools, and the increased attention to swarf collection generally have been prominent factors over many years, and the spread of pre-set or 'off-the-machine' tooling and automatic swarf-handling equipment are signs of the times heralding the disappearance of the conditions with which scrotal cancer is associated. The risk will be reduced when methods lessening the need for contact with oily components and machine parts are introduced, when efficient splash guards are provided and used and when oil fumes are extracted by an exhaust system. Jobs which entail prolonged contact with oil should be automated as far as possible.

Advice that a person who has been exposed to mineral oil and has developed a skin cancer should avoid further contact with the oil is often given but may not always be in the individual's best interests. Removal from his customary place of work may not prevent the development of growths from his past exposure and may induce a false sense of security. Furthermore, a skilled worker is unlikely to endure the loss of earnings that follows, and may soon take up his normal work elsewhere. It is sometimes more satisfactory for the individual to continue at his work under the regular medical supervision of his family doctor, or the hospital or works medical staff.

Use of solvent-refined and soluble oils Contact with a carcinogenic oil may be prevented by substitution of the oil or by a change of process. In the cotton industry, both factors have played a part. Mule spinning has now been largely superseded and under regulations introduced in 1953 any oil used in the process must be of animal or vegetable origin, or 'white oil' defined as 'a hydrocarbon oil of petroleum origin which has been drastically refined with sulphuric acid and conforms to the specified colour and the specified viscosity'.[3] In engineering, the use of neat and soluble cutting oils from which carcinogens have been removed by solvent extraction will obviate or greatly reduce the hazard. However, it will be many years before the statistics indicate conclusively whether the use of solvent-refined oils is completely successful in eliminating the occurrence of skin cancer. In the meantime the use of soluble oil (solvent-refined) where technically feasible, would also be an important step in the right direction, as would, in the future, the use of oil-free cutting fluids.

Overalls The wearing of oil-soaked clothing in continuous contact with the skin is to be deprecated and can be reduced by the provision of a regular laundry service for overalls. The wearing of impermeable protective aprons presents difficulties because the wearer may get hot and uncomfortable and the aprons may drip oil on to the feet. The application of an absorbent fabric over the impermeable material of the apron may be of value in preventing the dripping.

Washing facilities Good washing facilities are of great importance. Showers are provided for workers in pitch and tar and might well be considered where there is much contact with mineral oil. However, standards of cleanliness are not impugned: the evidence indicates that, in general, these are high in the workers affected.

Warnings The habit of putting oil-soaked wipers into the pocket should be discouraged.

There is a need to impress on workers that the oily wiper in the pocket can be the cause of something rather more serious than 'oil acne' on the thighs. Leaflets and posters exist which warn that mineral oil may cause cancer of the skin, of the scrotum in particular. It may be objected that they are not as forthright as they might be, or sufficiently intelligible to, for example, immigrant workers. There would seem to be scope for rethinking the presentation of the information they give and the warnings they sound and perhaps for the use of unambiguous coloured illustrations to supplement the wording. But posters are available and, accordingly, should be displayed.[4]

Medical examination

Legislation requires that workers in certain industries be medically examined at stated intervals—every six months for mule spinners. It is considered that if all those at risk and past and retired workers were examined regularly, then cancer of the skin would cause less disability and death than at present. Periodic examinations have been arranged for certain workers in a few engineering firms where there have been cases of scrotal cancer, and one large concern has started six-monthly checks for skin cancer on any part of the body of men exposed to mineral oil and is to collate an occupational history of each man which will be passed on when the man leaves. However, it has been pointed out that if every engineering firm were to institute frequent medical checks, doctors would be doing nothing else but seeking a disease which has killed about 10 engineering workers annually since 1963. And it is estimated that 80% of engineering companies do not have works doctors.

Malignant ulcers of the skin arising from contact with oil are notifiable under the Factories Act. The occupier of the factory in question must inform the District Inspector and the appointed factory doctor of any relevant case. But the patient's job designation may not tell a doctor sufficient for him to infer that there is an occupational risk, and an employer cannot be aware of a disease that an employee keeps to himself.

REFERENCES

1. Medical Research Council (1968). The Carcinogenic Action of Mineral Oils. A chemical and biological study S.R.S. 306 H.M.S.O.
2. A Symposium on Microbiology of Lubricating Oils (1967) Process Biochemistry, 2, 54.
3. Mule Spinning (Health) Special Regulations, 1953. Department of Employment and Productivity. H.M.S.O.
4. SHW 367 Cautionary notice on dermatitis. H.M.S.O. 3*d* net.
   SHW 295 Leaflet on 'Effects of Oil on the Skin'.
   SHW 397 'Effects of Mineral Oil on the Skin'.
   H.M. Factory Inspectorate (available on application).

8 FILTERS AND THE REMOVAL OF SWARF FROM CUTTING FLUIDS

INTRODUCTION

In these days of continually rising production costs and of increasing need to provide the highest possible finish on workpieces, filtration of the cutting fluids used on machine tools should be considered as essential to good machine-shop practice. From the point of view of economics alone, judicious filtration can extend the useful life of a cutting fluid by up to ten times and can often prolong the life of the cutting tool and result in savings in downtime; furthermore, improved surface finish resulting from adequate filtration will mean fewer rejects. Cutting-fluid life can also be extended considerably by use of a suitable filter, with the added advantage that problems of effluent disposal are reduced. Although many requirements might be satisfied by the use of a settling tank to remove a high proportion of the suspended solids, the use of an appropriate filter will reduce the need for cleaning the tank frequently and also the scope for breeding harmful bacteria.

The running cost of an appropriate filter will be most affected by the viscosity of the cutting fluid and the degree of filtration, the graph in Figure 8.1 showing how costs increase with increases in viscosity and finer degrees of filtration. It has been found in practice that cutting fluids of unnecessarily high viscosity are often used, thus increasing the running costs, but it is recognised that filtration down to micron size can often be beneficial and economic. It has been shown that, even with quite expensive equipment and taking running costs into account, expenditure can often be recovered within twelve months of installation.

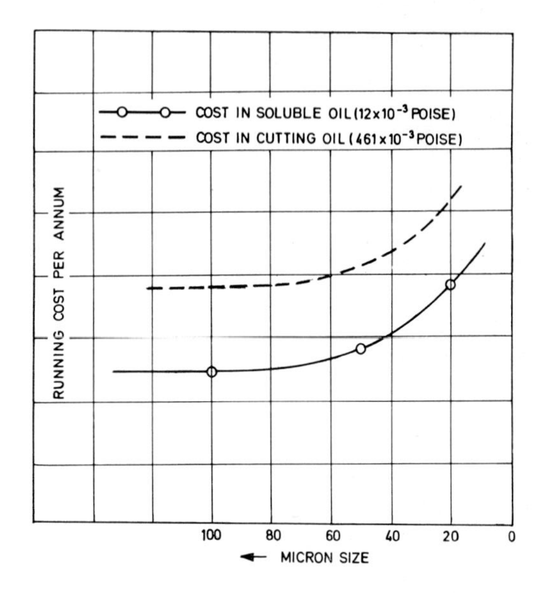

Fig 8.1 Graph showing increases in costs relative to increases in viscosity of cutting fluid and the degree of filtration

DEGREE AND RATE OF FILTRATION

Filtration is a process widely used in many industries and in each one a different level of performance may be required from the filtering appliances. It is difficult, therefore, to generalise on what is meant by degree of filtration but, in connection with machine tools, the term is taken to be a measure of the extent to which suspended solid particles of swarf are removed from a cutting fluid. To act as a guide only, and being aware of the potential errors, the following scale may be considered as reasonable for the clarification of cutting fluids:

1 Coarse filtration—the extraction of solids down to 250 micron
2 Medium filtration—the extraction of solids down to 50 micron
3 Fine filtration—the extraction of solids down to 10 micron
4 Extra-fine filtration—the extraction of solids down to 2 micron

Used cutting fluids may contain three types of contamination, these being:

1 swarf from machining processes;

2 bacterial growths;

3 plant dirt and other additions.

The swarf, and to some extent the plant dirt, can produce two serious conditions.

The first of these is poor surface finish for it has been shown that a fine surface finish of 4 micro-inches C.L.A. can deteriorate to 10 micro-inches C.L.A. in less than 120 seconds when a relatively small amount of swarf chips and plant dirt is introduced into the cutting fluid used in certain grinding operations. A summary of industrial experience on the maximum size of swarf particles permitted in the cutting fluids used in some machining operations has shown that the limit of particle size is approximately as follows:

	Machining operation	*Limit of particle size*
1	Surface and cylindrical grinding	0·5 micron
2	Thread grinding	3–5 micron
3	Honing	2–5 micron
4	Machine tapping	2–5 micron
5	Broaching	5–10 micron

The second serious condition is that tool life can be greatly decreased. Reductions of 15–20% in tool life on certain drilling applications due to dirty cutting fluids have been recorded, grinding-wheel life has been doubled by thorough chip removal, and other tools have been shown to suffer in varying degrees from swarf and other contaminants in the cutting fluid.

The degree of filtration achieved depends very much on the type of equipment used; with some filters it will remain constant whilst with those employing a filter medium it may vary with the thickness of the 'cake' of solids extracted and deposited on the medium. The cake, being of a porous nature, will act as a secondary filter medium and good cake formation is one measure of efficient filtration. Because of the characteristics of this type of filter it is often possible to specify a filter having a particle cut-off size which is 4–5 times greater than the average size of the particles required in the cutting fluid. For example, when a water-based, soluble cutting fluid is used in machining operations on steel, a filter having a particle cut-off size of between 50 and 150 micron is usually adequate to remove all but the finest particles and will give an average standard of clarity of about 25 micron. Decreasing the particle cut-off size to give an average particle size of, say, 5 micron will usually provide a significant improvement in tool life, but the additional cost of achieving this kind of performance must be carefully weighed against potential cost savings.

The filter cake has a very definite effect on the rate of filtration and as metallic swarf particles are under consideration, the following should be noted:

1 With precipitates of hard, non-deformable particles formed on well-defined filtering surfaces, the rate of filtration is proportional to the pressure and inversely proportional to the cake thickness.

2 Filter cakes of hard particles are usually uniform throughout and their character is unaffected by filtration pressure or solids concentration in the liquid to be filtered.

The finer the degree of filtration achieved, the greater will be the cleanliness of the cutting fluid, as removal of contaminants other than swarf particles renders the filtered fluid less likely to be polluted by bacteria.

FILTER MEDIA

Mention has been made of the use of filter media in some types of filtering appliance and it is necessary that the various forms and their uses be described. A filter medium may be regarded as a barrier to the passage of solid particles which allows the liquid portion of a mixture to proceed, and a filter can function successfully only if the correct medium is used.

The term filter medium is all-embracing and covers filters made from a variety of materials such as metal sheet, metallic wire, wool, jute, cotton, man-made fibres, sintered powders, etc., and prepared in the form of perforated sheet, a mesh or weave, or loosely compressed sheet. Filter media are manufactured under strict control so that the degree of porosity of each particular medium is known

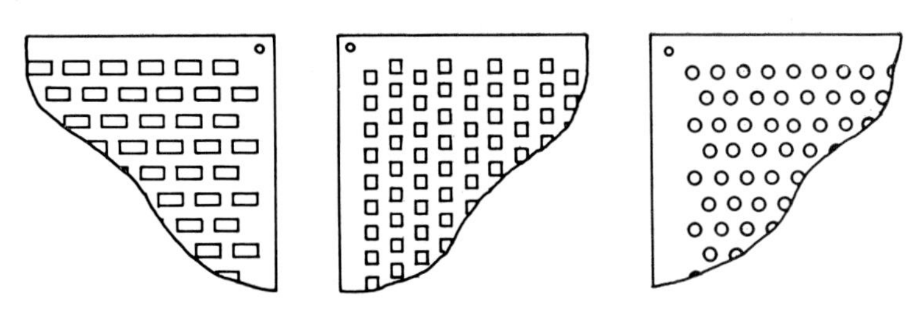

EXAMPLES OF PERFORATED METAL MEDIA

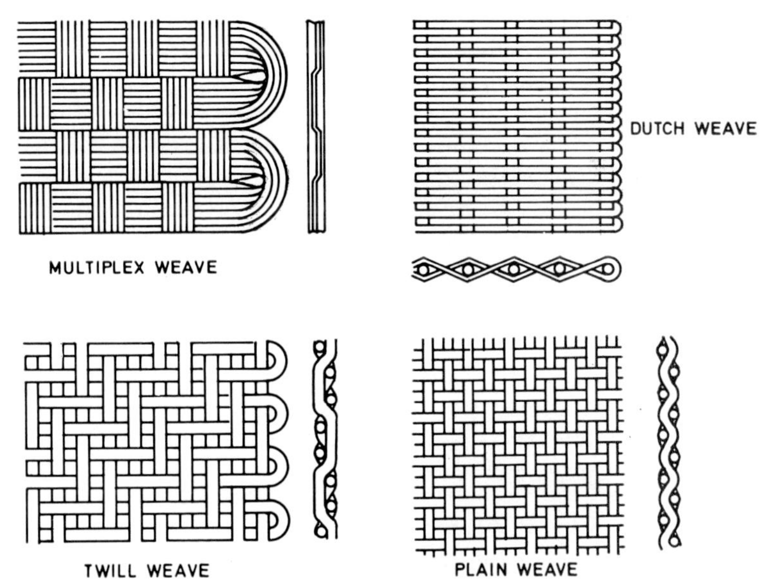

EXAMPLES OF WOVEN WIRE MEDIA

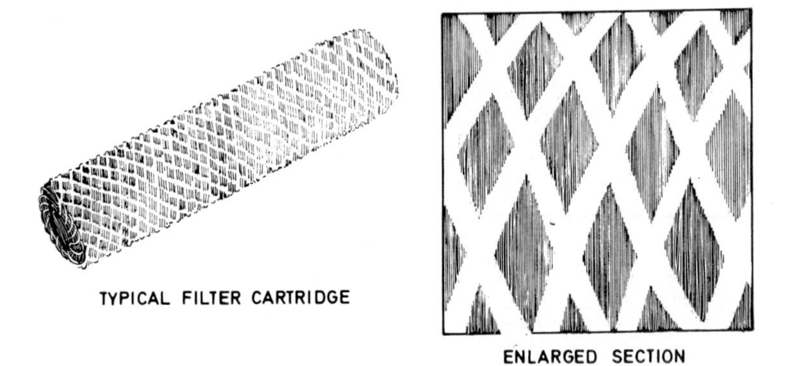

TYPICAL FILTER CARTRIDGE

ENLARGED SECTION

Fig 8.2 Typical example of filter media

and is maintained; this is also a measure of the permeability of a medium at the initial throughput of the solid-liquid mixture. Figure 8.2 shows examples of filter media.

Before a manufacturer can advise on the most suitable filter medium to be used for a specific process, he will consider both the particle count and the particle-size retention desired. Particle count is the information obtained by counting the particles of solid matter in groups based on relative size, and such data may be used both to engineer a filter and as a guide to the type of medium to be used. Particle-size retention is closely related to permeability or the ability of a medium to pass liquids under differential pressures through a material, the permeability measurement providing a convenient comparison for media and indicating the construction required for specific particle retention.

Filter media may be of the permanent type or the disposable type. In the case of the permanent type, for example hollow candles of compressed-paper or plastic discs, it will be necessary to remove the collected contaminants periodically and this operation may be carried out in a number of ways. Depending on the type of filter, contaminants may be removed by scraping, shaking, skimming, back-flushing, etc. Many of these cleaning operations may be carried out automatically when the back-pressure of the filter reaches a critical value. Disposable media, as the name implies, usually takes the form of a continuous moving membrane of filter material and this is thrown away with the contaminant after it has passed through the filtering zone. Examples of filters employing such media are given later in this chapter.

Another form of filter media may be described as 'loose solids' and these may consist of fibres of asbestos or cellulose, diatomaceous earth, expanded perlite, inactive carbon, adsorbent powders and coarse materials. Such media are often called 'filter aids' and they are used to pre-coat, i.e. create an artificial cake on certain normal media or to be used in a specially designed filter. The most commonly used material is diatomaceous earth which is a soft, earthy rock composed of the siliceous skeletons of small aquatic plants called diatoms (algae). It has a low specific gravity, is chemically inert and has the ability to maintain porosity in the filter cake. The filtering efficiency of such media is so great that they may

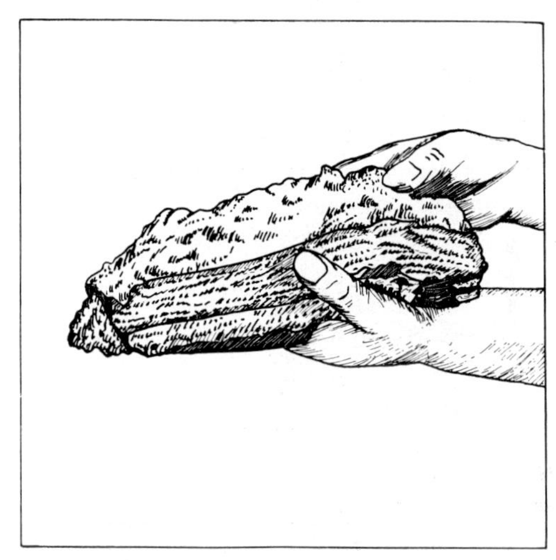

Fig 8.3 A good example of filter-cake formation

be dangerous if used with an emulsion type of cutting fluid, as not only will they remove the contaminants but also the active additives.

A good example of filter cake is shown in Figure 8.3. The cake holds together but it is still porous and spongy, permitting fine filtration through the swarf that collects on the filter media.

In Table 8.1 are given typical porosities of filter media.

TABLE 8.1 *Typical porosities of filter media*

Filter media	% *Free area*
Wedge wire screen	5–20
Perforated sheet	20
Wire mesh: Twill weave	15–25
Square	30–35
Porous plastics, metals and ceramics	30–50
Crude kieselguhr	50–60
Porous ceramic, special	70
Membranes, plastic foam	80
Asbestos/cellulose sheets	80
Refined filter aids (diatomaceous earth, expanded perlite)	80–90
Paper	60–95
Plastic foam	97
Nylon polypropylene monofilament	15–40

Table 8.2 offers a rough guide to the general classification of many types of filter media on the basis of rigidity.

TABLE 8.2 *General classification of filter media on the basis of rigidity*

Type	*Examples*	*Minimum size of particle arrested Micron*
Solid fabrications	Scalloped washers, wire-wound tubes	5
Rigid porous media	Ceramics and stoneware	1
	Sintered metal	3
Metal sheets	Perforated	100
	Woven wire	5
Porous plastics	Pads, sheets, etc.	3
	Membranes	0·005
Woven fabrics	Cloths of natural and synthetic fibres	10
Cartridges	Yarn-wound spools, graded fibres	2
Non-woven sheets	Felts, lap, etc.	10
	Paper-cellulose	5
	Paper-glass	2
	Sheets and mats	0·5
Loose solids	Fibres, e.g. asbestos and cellulose	Submicron
	Powders such as diatomaceous earth	Submicron
	Expanded perlite	Submicron
	Inactive carbon	Submicron
	Adsorbent powders	Submicron
	Coarse materials	Submicron

CLEAN CUTTING FLUID AND CUTTING-TOOL LIFE

Mention has previously been made of the economic advantages that can accrue from the use of cutting fluids from which all contaminants down to small micron sizes have been removed. In this section, it is the intention to offer the apparent reasons why clean cutting fluids, regularly used, will prolong the life of the cutting edges of cutting tools.

A grinding wheel may be considered as a multitude of single-point cutting tools in the form of grit particles bonded together by a suitable agent. During most grinding operations, swarf is produced from the workpiece in the form of very small metallic particles that are washed away by the cutting fluid but, during the process, it is inevitable that a certain amount of wear of the grinding wheel also takes place. Because of this wear, the cutting fluid will contain abrasive grit and a proportion of the bonding agent as well as the metallic particles, and all will be carried forward to the cutting-fluid tank.

Most grinding machines are equipped with some type of magnetic separator in the cutting-fluid tank and these are usually very efficient in removing the metallic swarf and, indeed, some of the other contaminants that adhere to the metallic particles, but the majority of the grinding-wheel debris will not be extracted. Although a certain proportion of this debris will settle-out in the tank, it is obvious that the majority is likely to be re-circulated through the cutting-fluid system and be responsible for:

1 'loading' or 'glazing' of the face of the grinding wheel necessitating a dressing operation;

2 inferior surface finish due to the trapping of particles between the workpiece and the grinding wheel.

It follows, therefore, that the fewer the contaminants in the cutting fluid the longer the life of the cutting face of the grinding wheel between dressings and the better the surface finish of the workpiece; the answer lies in efficient filtration of the cutting fluid. Experimental work has been carried out to determine the effects of contaminated cutting fluids on surface finish and Figure 8.4 clearly illustrates the results obtained.

The effect of the use of uncontaminated cutting fluid on the life of the cutting edge of a single-point cutting tool is somewhat more difficult to determine, partly because of the lack of research on this aspect of cutting tools and partly due to a lingering doubt as to how cutting fluids reach the area where the cutting tool makes contact with the workpiece and where the chip is produced.

A great deal of investigation has been carried out in an effort to resolve this latter problem but it still remains a matter of conjecture. Some workers maintain that the chip separates from the workpiece a little ahead of the tip of the cutting tool and suggest that the small space so formed between tool, workpiece and chip forms a partial vacuum into which the cutting fluid is drawn, whilst other workers contend that capillary action and vibration play the major part in feeding the fluid to the cutting zone. It is probable that both theories are more or less correct according to the actual cutting conditions, and that it is by a combination of forces that the cutting fluid is able to enter the vital area, as shown in Figure 8.5.

Along with the swarf chips that are produced during a normal cutting operation there will always be a number of very fine solid particles that become suspended in the cutting fluid and which cannot be removed except by fine filtration, These fine particles are circulated with the cutting fluid and, during the actual cutting process, they become trapped between the tool and the workpiece or between the tool and the chip where they are subjected to very large frictional forces. Because of the friction, there is a very considerable increase in temperature at the tool tip where penetration by the cutting fluid is most difficult and, as a result, some of the fine particles may become welded to the tool tip and cause rapid deterioration of the cutting edge, i.e. there is a tendency for the cutting edge to become built up. As the ability to cut decreases rapidly with increase in height of the built-up edge, it becomes obvious that the longer the build-up can be delayed the longer will be the life of the cutting edge of the tool between re-grinds.

The fine metallic particles that are being considered may be only a few micron in size but if their presence in the cutting fluid is

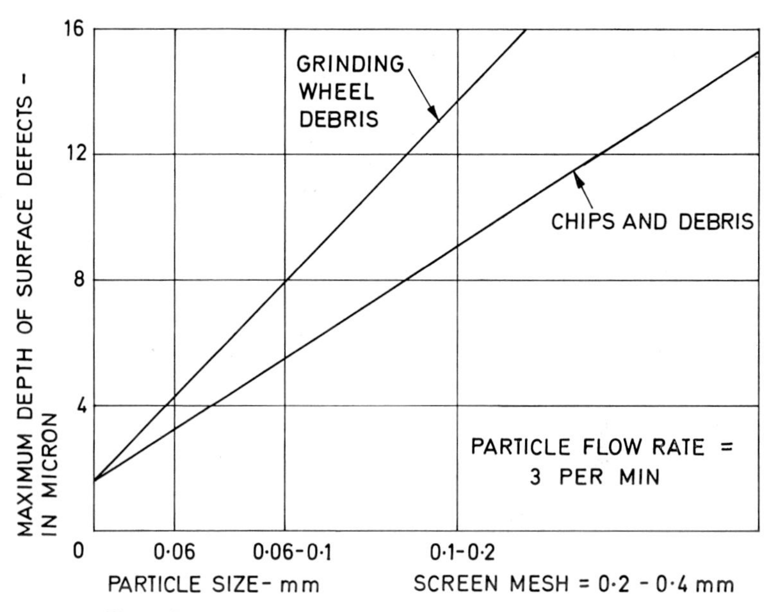

Fig 8.4 The effect of particle size on workpiece surface finish

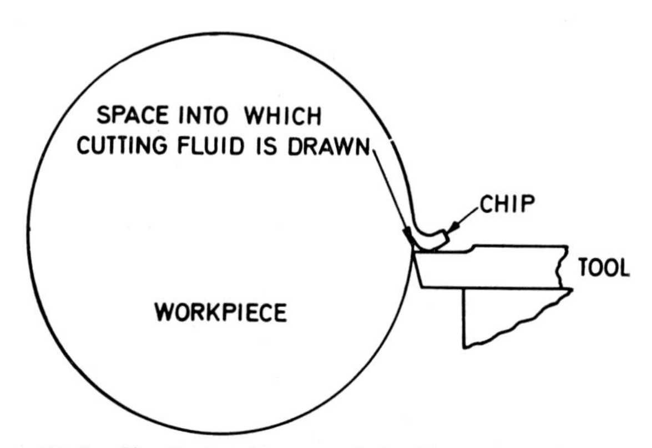

Fig 8.5 Sketch showing space formed between tool and workpiece

economically undesirable they can be removed by fine filtration and a clarified fluid obtained. The savings in cost can be considerable as all the following may be regarded as benefits gained:

1 longer life of cutting edge between re-grinds;
2 lower consumption of tool steel or carbide tips;
3 less toolroom time on re-grinding;
4 less machine down-time;
5 longer useful life of the cutting fluid;
6 less time on cleaning-out of cutting-fluid systems.

SELECTION OF A FILTER

Before selecting the type of filter to be used, it is necessary to be aware of the requirements so that the most suitable and economic type can be offered. It must be expected that the finer the degree of filtration and the higher the viscosity of the cutting fluid, then the greater the initial outlay and running costs. Whilst fine filtration normally justifies the extra expense, attention to the viscosity of the cutting fluid, often higher than required, will usually produce a saving.

Because of the various possibilities, considerable investigation will usually be found necessary before the equipment manufacturer is able to offer his advice but information on the following will certainly influence the selection:

1 What degree of filtration is required?
2 What type and viscosity of cutting fluid is to be treated?
3 What rate of stock removal is anticipated, i.e. what will be the particle content of the fluid?
4 What volume of cutting fluid will need treating per hour?

Other questions to be answered at this stage are:

5 What is the least expensive process for removing the contaminant, i.e. would 'settling-out' be of assistance?

6 What precautions in locating a filter are necessary so as to guarantee maximum efficiency and constant flow?

7 What construction requirements of the filter should be considered, such as size, rigidity, pressure resistance, etc.?

8 Will the filter be too stringent and remove from the cutting fluid the desirable additives as well as the undesirable contaminants?

TYPES OF FILTER

In this section the main types of filter are described, with notes on the method of operation and performance.

MAGNETIC SEPARATORS

Automatic drum type

(See Figure 8.6) This type of equipment is suitable for separating ferrous particles from cutting fluid, the main feature being a drum

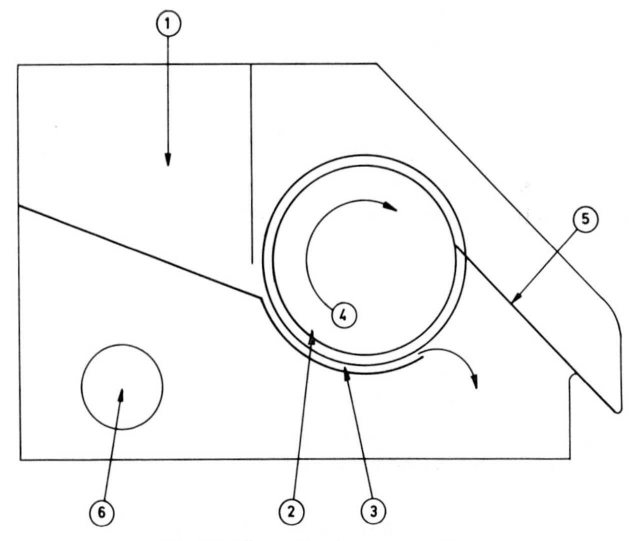

Fig 8.6 Magnetic-drum separator

1. Contaminated-coolant inlet
2. Magnetic drum
3. Flow path
4. Direction of rotation
5. Swarf removed from drum by scraper
6. Clean-coolant outlet

containing permanent magnets arranged with suitable pole plates to give a high-intensity magnetic field. The separation is essentially automatic, depositing swarf into a waste container.

Method of operation Contaminated cutting fluid from a workpiece is fed to the separator inlet and, due to the construction of the bodywork, forced to pass through the high-intensity magnetic field. The drum revolving slowly extracts the ferrous particles together with a percentage of entrained non-ferrous swarf. The waste is then lifted to a higher level where a doctor blade removes the swarf from the drum.

Application For removing ferrous swarf created during operations on grinding machines, drilling machines, gear-hobbing machines, etc. Suitable for water-soluble or low-viscosity oil cutting fluids.

Performance Flow rates from 1 g.p.m. (5 $l\,min^{-1}$) to 100 g.p.m. (720 $l\,min^{-1}$) dependent on application. Individual sizes can be grouped together for increased flow rates.

The separation efficiency is dependent on both the flow rate and viscosity. Figure 8.7 shows performance curves for this type of separator. In the tests referred to, two types of contaminant were used, one containing 23 % grit and the other 7 % grit. Although the grit is non-magnetic the over-all efficiency is still 98 %, indicating that the steel parts attach themselves to the grit and thereby assist in their removal.

Automatic conveyor type

(See Figure 8.8) Normally used for the removal of large chips from cutting fluid. Two types are available. In one, permanent magnet assemblies move under a non-magnetic skin; the other incorporates a belt moving over stationary magnets.

Method of operation Cutting fluid containing ferrous chips is delivered to the inlet hopper of the separator and forced to pass through the magnetic field provided by the permanent magnet system of the conveyor. The chips are then attracted by the con-

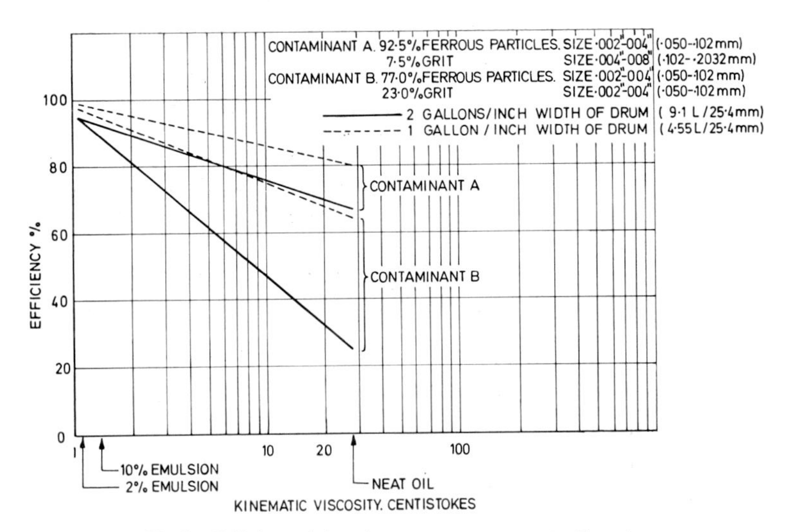

Fig 8.7 Efficiency/viscosity curves – magnetic filtration

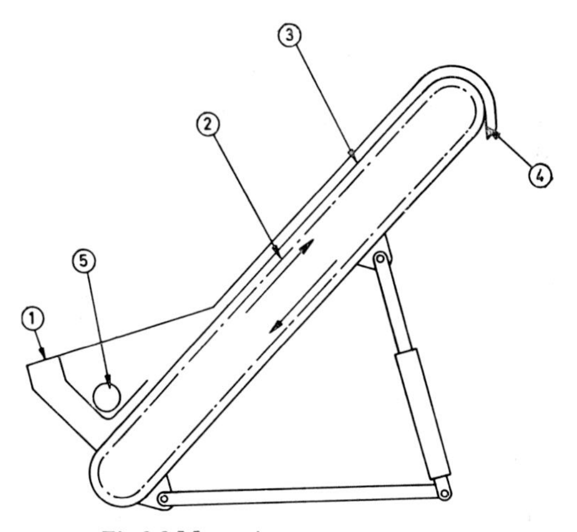

Fig 8.8 Magnetic-conveyor separator

1. Contaminated-coolant inlet
2. Magnetic assembly travelling in direction of arrow
3. Non-magnetic skin
4. Swarf discharge
5. Clean-coolant outlet

veyor and deposited at high level, whilst the cleaned cutting fluid spills back to the cutting-fluid tank.

Application Ferrous chips generated by drilling, hobbing or automatic machines; ideally suited for incorporating into machine tools with integral sumps too low for other forms of separation. Suitable for water-based or neat-oil cutting fluid.

Performance Flow rates up to 60 g.p.m. (270 l min^{-1}) removing up to 5 lb per min (2·3 kg min^{-1}) of chip.

Pipe-line type

(See Figure 8.9) This type of clarifier should not be considered for the removal of bulk swarf but as a final cleaning operation before cutting fluid is delivered to the workpiece.

Method of operation Cutting fluid is pumped through a housing containing magnetic assemblies of either permanent or electromagnetic type. The flow is normally through the energised assembly thus leaving ferrous particles within the trapping area. The cutting fluid, which is under pressure, may be delivered direct to the work or to a storage tank for re-circulation.

In the case of the electromagnetic type cleaning may be carried out by de-energising and flushing through. In the case of permanent magnets, removal of the trapping area from the core will de-energise it for cleaning.

Application For the final clearing of ferrous particles from water-based or oil cutting fluids.

Performance Flow rates of up to 1,700 g.p.m. (7,650 l min^{-1}).

A definite cut-off point cannot be stated but a degree of separation down to micron-size ferrous particles can be expected.

GRAVITY FILTERS (see Figure 8.10)

Gravity filters of the type illustrated usually make use of an endless belt of mesh shaped to form a trough to support the filter medium

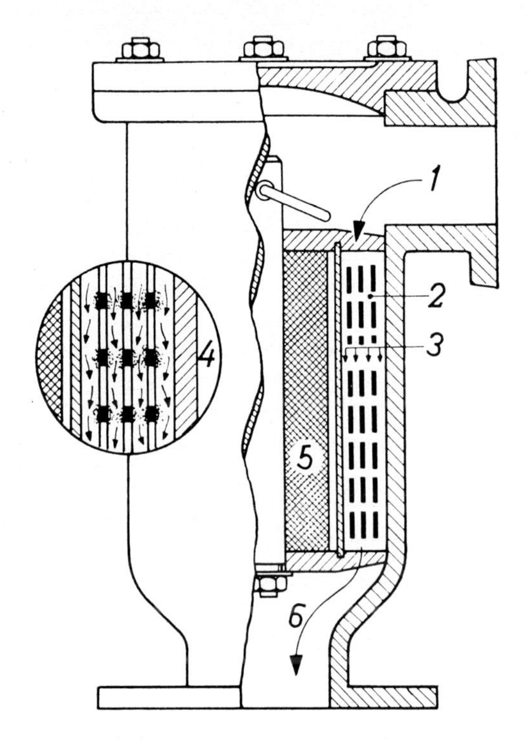

1 = dirty oil or coolant inlet
2 = magnetized iron filter cage
3 = free flow of liquid passing through filter cage
4 = ferrous contamination collects only in gaps between filter segments (see insert)
5 = powerful permanent magnet
6 = clean-oil or coolant outlet

Fig 8.9 Magnetic separator for use in pipe lines

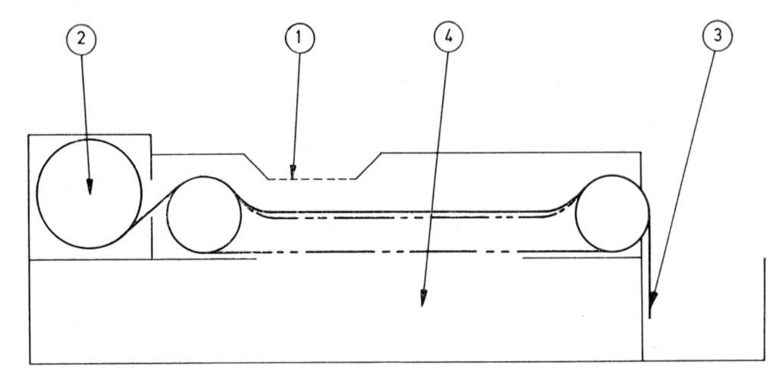

Fig 8.10 Gravity filter

1. Contaminated-coolant inlet
2. Filter media roll
3. Disposable filter media and solids
4. Clean-coolant tank

and often incorporate a tank in the base of the unit to retain the cleaned cutting fluid.

Method of operation Contaminated cutting fluid passes into the trough supporting the filter medium. Particles of swarf are arrested by the filter medium and the clean cutting fluid passes through to a clean storage tank from which it can be re-circulated to the machining zone. As the filter medium becomes loaded with swarf particles, so the level of dirty cutting fluid rises. There are several ways of sensing when the filter is blocked sufficiently to need replacing, i.e. liquid-level switch, pressure switch, weight, timer, etc.

When the pre-set control point has been reached an electric motor actuates the drive to the mesh belt which brings forward a fresh length of filter medium from a roll positioned at the rear of the equipment. The used filter medium together with the swarf attached are then deposited in a bin specially designed to accept the waste material.

The filter medium is disposable, normally made of non-woven bonded fibre, although woven materials are available.

Application Suitable for most machine-tool operations using water-based or low-viscosity cutting fluids.

Performance The degree of filtration is dependent on the grade of filter medium used; a nominal 15 micron can be obtained with a medium in its clean state but this improves progressively as contamination builds up. The cost of replacement media is dependent on viscosity and concentration of swarf.

Flow rates from 3 g.p.m. (14 l min^{-1}) to 450 g.p.m. (2,000 l min^{-1}) of water-soluble fluid. Through filter area 1·25 sq. ft (0·1 m^2) to 88 sq ft (8 m^2). Floor areas 5 sq ft (0·45 m^2) to 200 sq ft (18 m^2).

COMBINED MAGNETIC AND GRAVITY FILTER (see Figure 8.11)

As the title implies, this is a combination of a magnetic and a gravity filter.

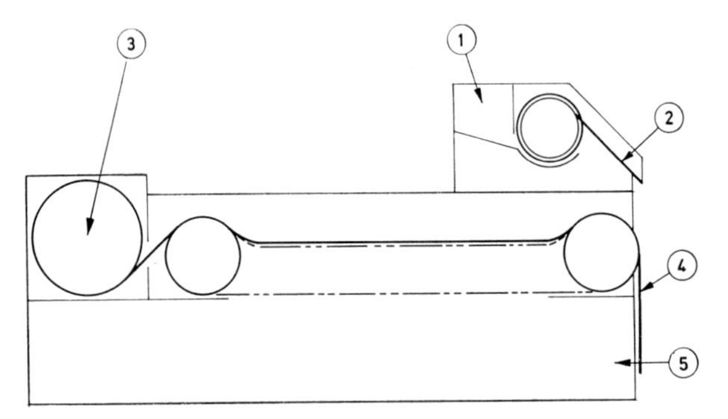

Fig 8.11 Combined magnetic and gravity filter

1. Contaminated-coolant inlet
2. Ferrous-swarf disposal point
3. Filter media roll
4. Disposable filter media and remaining contaminants
5. Clean-coolant tank

Method of operation Contaminated cutting fluid is first fed through the magnetic-drum separator, thus removing the bulk of the ferrous swarf and relieving the filter medium on the gravity unit. From the first stage, the cutting fluid passes through the medium to the storage tank for re-circulation.

Application For heavy grinding operations using water-based or low-viscosity cutting fluids. Special attention must be paid to the discharge height of cutting fluid from the machine tool, bearing in mind that one clarifier is mounted on the other.

Performance The final degree of filtration will be dependent on the grade of filter medium used; a nominal 15 micron can be obtained with media in the clean state but this improves progressively as contamination builds up. The cost of replacement medium is dependent on cutting-fluid viscosity and concentration of contaminant.

Flow rates 8–100 g.p.m. (36–450 l min^{-1}). Filter areas 2·3–35 sq ft (0·2–3·2 m^2). Floor areas 9·5–75 sq ft (0·8–6·8 m^2).

VACUUM FILTERS (see Figure 8.12)

This is an automatic unit which, by use of a pump creating a vacuum, uses atmospheric pressure to force cutting fluid through a disposable filter medium. There are two forms—the horizontal and the rotary drum type.

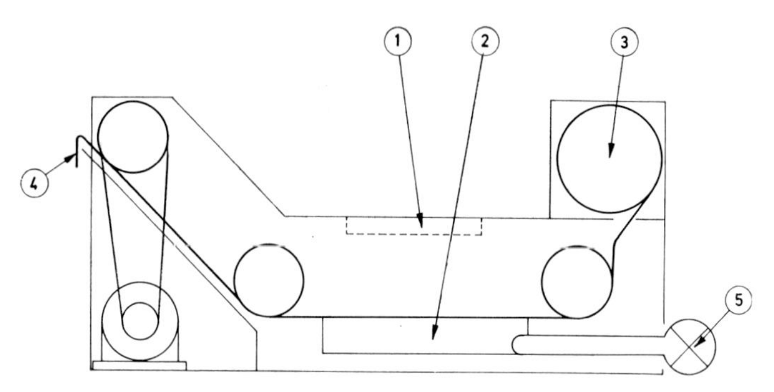

Fig 8.12 A horizontal-type vacuum filter

1. Contaminated-coolant inlet
2. Vacuum chamber
3. Filter media roll
4. Disposable filter media and solids
5. Clean coolant outlet

Horizontal type

Method of operation The unit consists basically of a tank divided into two sections, the upper part being open to atmosphere, the lower containing a vacuum chamber which is connected to the suction pump. Contaminated cutting fluid is fed by gravity or pumped into the upper chamber, to be forced by atmospheric pressure on the surface of the media through the vacuum chamber into a buffer or storage tank for re-circulation.

A flight conveyor is in the upper compartment. As the filter medium becomes blocked, this is detected by a level switch or pressure differential and the suction pump is stopped, the vacuum broken and the conveyor drive is actuated, bringing forward fresh medium from the roll positioned at the rear of the clarifier, the contaminated medium being deposited automatically in a waste container at the front end.

As an alternative, the conveyor movement may be on a fixed time cycle interlocked with the machine-tool operation.

Application To individual machine tools where cutting-fluid flow rate is high and floor space at a premium. Ideally suitable for centralised systems, serving a number of machine tools through one or more clarifiers. Can be used with water-based or neat-oil cutting fluids.

Performance Capable of flow rates from 13 to 1,250 g.p.m. dependent on cutting fluid viscosity and concentration of contaminant. By virtue of the fact that the contamination acts as a pre-coat on the media, fine filtration down to 5 micron particles can be obtained.

Filtration areas from 3·3 sq ft to 52 sq ft (0·3 to 5 m^2). Floor areas 12·6 sq ft to 126·5 sq ft (1 to 12 m^2).

Rotary-drum type **(see Figure 8.13)**

This type of filter uses a rotary drum which is evacuated, thus causing the cutting fluid to pass through a filter medium situated on the periphery of the drum.

Method of operation This type of filter consists basically of a tank in which a revolving drum is mounted, the contaminated cutting fluid being delivered into the tank by a pump. During operation,

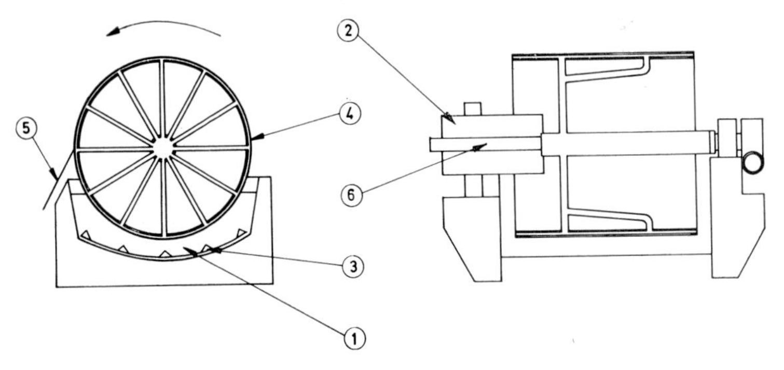

Fig 8.13 A rotary-drum type of vacuum filter

1. Contaminated-coolant inlet
2. Valve
3. Agitator
4. Drum rotating in direction of arrow
5. Scraper depositing contaminants
6. Clean-coolant outlet

the filter drum is partially submerged, sometimes so deeply as to cover 60% of the surface to which the filter medium is attached, but an adjustable overflow valve is provided on the tank so that any excess of cutting fluid is returned to the surge tank on the suction side of the pump. In order to prevent the solid contaminant from settling on the bottom of the tank the cutting fluid is kept in motion by an agitator.

The filter is equipped with a vacuum pump which reduces the pressure inside the drum via a connection made through a centrally placed hollow shaft.

When the vacuum pump is operating, the liquid portion of the contaminated cutting fluid is drawn through the medium with which the drum is covered and the solid contaminant is retained on the surface of the medium. As the drum rotates, the solid contaminants are lifted clear of the fluid and they may be removed from the surface of the medium by a scraper blade or by jets from cleaning nozzles.

When a high degree of clarification of the cutting fluid is required, this type of filter may be used with a suitable pre-coat applied to the outer surface of the drum.

Application This filter is used only for grinding-machine cutting fluids where there are fine solids. It is not economical unless included in a central system but is useful to recover solids where necessary. It can be used with water or oil-based cutting fluids.

Performance Capable of dealing with from 20 g.p.m. (90 l min^{-1}) to 2,500 g.p.m. (11,365 l min^{-1}) dependent on cutting-fluid viscosity and solid contents. Since pre-coat can be used with this type of filter a fine degree of filtration can be achieved.

Filtration area from 5 sq ft to 100 sq ft (0·5–10 m^2). Floor area 30 sq ft to 400 sq ft (3 to 38 m^2).

PRESSURE FILTERS

Pressure filters are dependent on a pump forcing cutting fluid through a suitable filtering material which may be a disposable or permanent medium, wire screen or pre-coat. The cleaned cutting fluid may be returned under pressure direct to the work or to a storage tank for re-circulation.

Examples of the various types available are given below.

Bag type (see Figure 8.14)

Method of operation Contaminated cutting fluid is fed to a tank containing a pump capable of handling swarf and abrasive grit. The pump forces the dirty cutting fluid into a cylinder containing a permanent or re-usable filter element, usually in the form of a bag or cone.

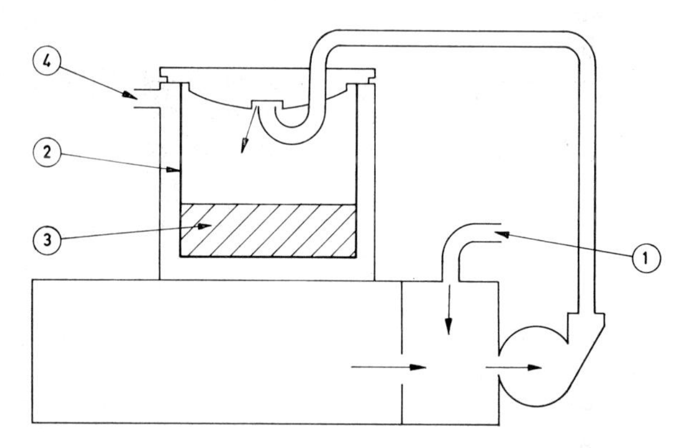

Fig 8.14 A bag-type pressure filter

1. Contaminated-coolant inlet
2. Filter bag supported by perforated basket
3. Sludge
4. Clean-coolant under pressure outlet

The bag retains the swarf, allowing clean cutting fluid to be returned direct to the work or storage tank. As the element becomes blocked, so the rate of cutting-fluid flow falls, indicating that cleaning is required.

By the use of the duplex or multi-pot type of filter, servicing can be carried out without interruption of flow.

Some pressure filters are available using a permanent medium which allows cleaning automatically on a time cycle or pressure differential.

Application To most grinding and cutting machines, using soluble oil/water, light neat oils and paraffin cutting fluid.

Performance Standard equipment up to 100 g.p.m. (450 l min^{-1}) of soluble oil/water. Degree of filtration dependent on the type of element used. Generally 15-micron filtration is achieved, reducing to 5 micron upon build-up of sludge forming a self coat with the element.

***Drum type* (see Figure 8.15)**

Method of operation The drum filter consists of a pressure-tight casing having a cylindrical top section and conical base; the drum mounted on a shaft revolves in the cylindrical part of the casing and a cleaning nozzle is positioned close to the surface of the drum.

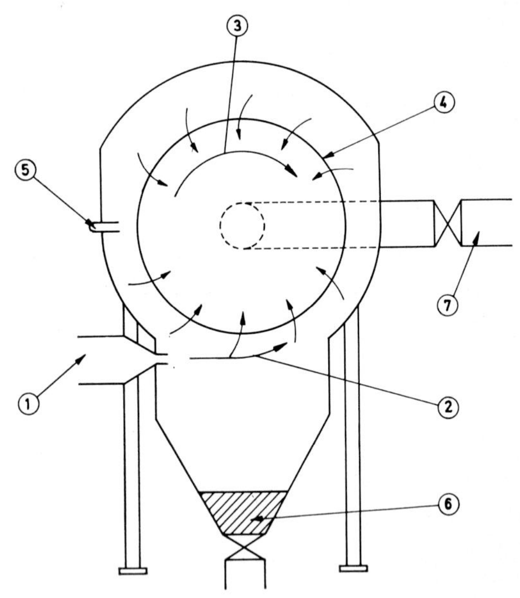

Fig 8.15 A drum-type pressure filter

1. Contaminated-coolant inlet
2. Flow path
3. Direction of rotation
4. Drum with filter medium covering
5. Cleaning nozzle
6. Contamination discharged through valve
7. Clean-coolant outlet

The drum has closed ends and a perforated surface covered with mesh over which is stretched a filter band.

Cutting fluid is pumped into the casing and passes through the medium on the drum, being discharged through a pipe from the centre of the drum. Heavy contamination falls to the lower conical section; contamination held by the medium is removed by the action of the cleaning nozzle.

Fine swarf is removed by a nozzle and heavy swarf is removed through a valve at the base of the cone. The filter is fully automatic in operation.

Application A range of sizes allows this type of clarifier to be applied to individual machine tools or on a centralised system.

Performance The degree of filtration is dependent on grade of media used. An average standard of filtration of less than 10 micron can be obtained using the appropriate filter band.

Filter areas from 4 sq ft to 40 sq ft (0·4 – 8 m^2). Flow rates up to 2,500 g.p.m. (11,250 l min^{-1}).

Pipe-line type

A large variety of in-line pressure clarifiers is available. In all cases, cutting fluid must be pumped under pressure into the filter to pass through the medium direct to work. A continuous flow of cutting fluid may be maintained by using a duplex unit or one of the self-cleaning types. These consist of a housing containing one or a number of filtering elements which can be of mesh screen, edge strainers, wire-wound or leaf filters and, in many cases, pre-coat can be applied for greater filtration efficiency. They are obtainable as single or duplex units manually or automatically cleaned. Figure 8.16 shows one of the variations available.

Performance Extremely fine filtration can be achieved with use of pre-coats but, with mechanical or plate-edge type, down to 10 micron only can be obtained.

Flow rates of up to 10,000 g.p.m. (45,000 l min^{-1}) depending on

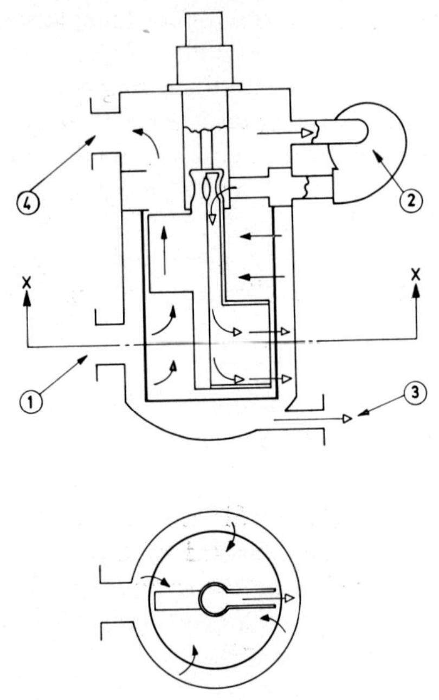

Fig 8.16 A pipe-line type of pressure filter

1. Contaminated-coolant inlet
2. Backwash pump
3. Flush to tank
4. Clean-coolant outlet

Coolant flow path ——▶
Backwash flow path ——▷

the viscosity of the cutting fluid and the permissible pressure drop across the filter.

Leaf type (see Figure 8.17)

Method of operation The horizontal leaf filter consists of a pressurised casing having a stack with horizontal filter leaves mounted on a common central shaft.

The cutting fluid is pumped into the main casing and passes through the filter nest and into the central shaft, which is hollow, from where it is discharged. Contamination builds up on the top surface of the filter screens, and when they require discharging the central shaft is rotated, thus throwing the solids into the main casing and out through the slurry discharge. The contamination can be air-dried to produce a dry residue if this is required.

Application A range of sizes allows this type of clarifier to be applied to large individual machine tools or to a centralised system.

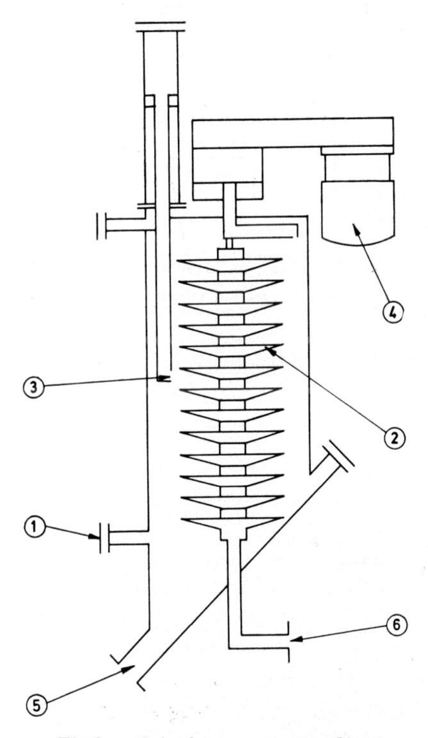

Fig 8.17 A leaf-type pressure filter

1. Contaminated-coolant inlet
2. Filter screen
3. Compressed air blow off
4. Motor to rotate filter screen
5. Contaminant discharge
6. Clean-coolant outlet

Performance The degree of filtration depends upon the grade of media used, these being available from 40 micron to 130 micron in stainless steel, cotton, polypropylene, teflon or nylon.

By pre-coating, extremely fine filtration can be obtained.

Filter areas from 1 sq m to 60 sq m. Flow rates up to 66,000 g.p.h.

***Flat-bed pressure clarifier* (see Figure 8.18)**

Method of operation The flat-bed filter comprises two shells, the upper one being movable and the lower fixed. Filter medium, either permanent in the form of an endless belt or a roll of disposable material, is held between the two shells, and sealing effected by pneumatic cylinders.

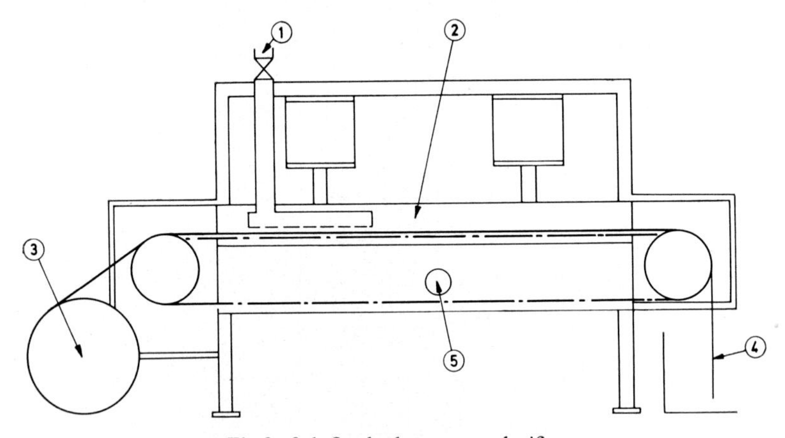

Fig 8.18 A flat-bed pressure clarifier

1. Contaminated-coolant inlet
2. Top shell (movable)
3. Filter media roll
4. Disposable filter media and solids
5. Clean-coolant outlet

Contaminated cutting fluid is pumped into the top shell and passes through the medium, leaving swarf deposited on the surface. The continued build-up of contaminant offers an increasing resistance to flow which causes a resultant rise in pressure within the top shell. On reaching a pre-set limit, the pressure rise initiates the automatic cleaning cycle.

During cleaning, the inlet valve is closed. Low-pressure compressed air is admitted to the top shell to blow the remaining liquid through the dirt cake until dry. The air supply is closed, the top shell raised and conveyor set in motion to deposit waste in a container and bring forward fresh medium. The top shell then closes and the inlet valve opens to allow filtration to continue.

Application To almost all machining operations on cutting and grinding machines. Can be applied to individual machines or to centralised systems.

Performance A fine degree of filtration can be achieved by use of pre-coats.

Flow rates of up to 1,250 g.p.m. (5,600 l min^{-1}). Filter areas 2·6 sq ft to 52 sq ft (0·2 – 5 m^2). Floor areas 6 sq ft to 130 sq ft (0·5 – 11 m^2).

FLOTATION SYSTEMS (see Figure 8.19)

Method of operation

When the contaminated cutting fluid enters the clarifier through the header pipe all the heavier swarf particles settle to the bottom of the chamber whilst the lighter particles which remain in suspension flow through the tank. Flotation of these lighter solids is accomplished by the introduction of millions of microscopic air bubbles into the cutting fluid. As these bubbles rise, they attach

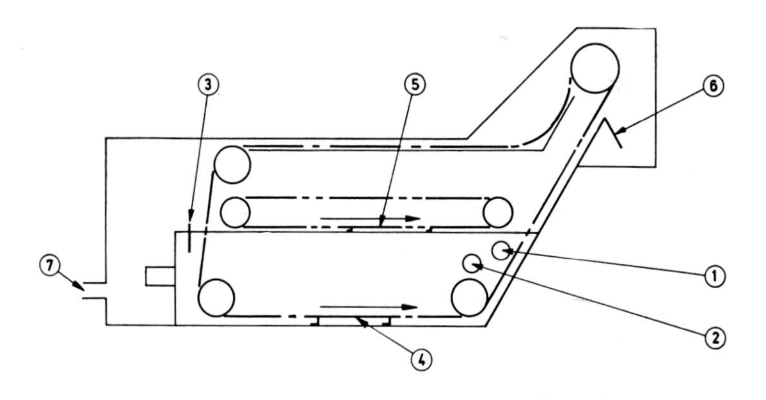

Fig 8.19 A flotation-type clarifier

1. Contaminated-coolant inlet
2. Recycle header pipe
3. Baffle to stop carry over of lighter particles
4. Scraping mechanism
5. Skimming mechanism
6. Sludge deposited into bin
7. Clean-coolant outlet

themselves to the particles in suspension and carry them to the surface for removal.

An important principle of the flotation system is the introduction of air into the re-circulated cutting fluid, and blending the re-circulated flow with the raw flow. The recycle liquid is pumped at a relatively high pressure to a suitable pressure-retention tank where air is introduced. While flowing through the pressure tank, the maximum possible quantity of air is dissolved in the recycle stream. The air-saturated stream is then fed back into the tank through a header and blends with the raw flow.

As the pressurised flow is restored to atmospheric pressure, millions of microscopic bubbles are formed. These air bubbles attach themselves to suspended particles in the flow and lift the particles to the surface for effective removal.

The removal of the lighter particles from the surface of the liquid is accomplished by the skimming mechanism. This conveys the floating solids to the ramp end of the tank. Here they are removed by the main scraper mechanism and, together with the large swarf particles scraped from the tank floor, are drawn up the ramp and discharged into bins.

Application Can generally be applied to all machine-tool operations involving water-based cutting fluids, particularly on centralised cutting-fluid systems.

Performance Unlimited flow rates. Degree of filtration to 16 micron.

PRE-COAT FILTERS

Whilst some of the foregoing equipment can be pre-coated to improve performance, the term 'pre-coat filter' usually refers to a filter specifically designed to use the process of pre-coating with diatomaceous earth and such filters cannot normally be used without pre-coating.

The filter consists of a multiplicity of septa in the form of either plates, discs or tubes. Prior to passing contaminated cutting fluid through the filter, a slurry of the liquid and diatomaceous earth is pumped through, the diatomaceous earth being deposited on the septa in the form of a uniform layer, hence the term 'pre-coat'.

The characteristics of diatomaceous earth are such that this coating forms a porous bed capable of retaining particles down to approximately 5 micron or less on the initial passing of the fluid. The main use of the filter is under conditions where this degree of cleanliness is necessary from the outset.

The filter is normally cleaned by back-flushing, sometimes associated with a mechanical process, such as spinning, where the septa consist of discs.

This type of filter becomes economically viable where it is necessary to filter at a high flow rate combined with a relatively high viscosity.

9 THE USE OF CENTRIFUGAL SEPARATORS AND HYDROCYCLONES IN THE REMOVAL OF SWARF FROM CUTTING FLUIDS

INTRODUCTION

As an alternative or in addition to the use of filters for the removal of solids from liquids, centrifugal separators and hydrocyclones may also be used.

Unlike the process of filtration wherein the separation of solids from liquids is accomplished by the intervention of a material medium which arrests the progress of the solids in the solid/liquid mixture whilst allowing the liquid to continue its flow, in centrifugal separators and hydrocyclones the separation is made possible by the action of centrifugal force, hence the name 'centrifuge'.

Centrifugation is capable of separating solids from liquids or of classifying granular material, e.g. swarf chips, according to particle size. In certain cases the separation of heterogeneous liquid mixtures is also possible. In the context of separating swarf from cutting fluids centrifugal separators are capable of removing a broad range of particle sizes and the limit is approached when the particle size is about one micron.

Apart from being a speedy method of bringing about separation, the following factors have an important bearing on the reasons why centrifugal separators are particularly suitable for processing swarf and cutting fluids following machining operations:

1 As an economic measure, to recover the cutting fluid in a sufficiently clarified condition to make it suitable for direct re-use.

2 To separate and dry the swarf particles, so that they are in the

best saleable condition, thus commanding the highest prices from the scrap-metal merchant.

3 Because of the degree of clarification that can be achieved in the cutting fluids, surface finish of the workpiece can be maintained at a satisfactory level, tool life between re-grinds may be extended, the life of machine-tool slideways and bearings is prolonged and bacterial action is minimised.

4 Some centrifugal separators are capable of removing water from neat cutting oils, separating tramp oils from soluble-oil emulsions, and in some cases it is claimed that soluble oil may be separated from neat cutting oils.

5 Although it is suggested that the recovery of soluble-oil emulsions from swarf is not an economical proposition unless swarf is produced in great quantity, a centrifugal separator will clean such cutting fluid from the swarf particles. In view of the legislation now affecting the disposal of effluents and the attitude of local authorities, it may become necessary to remove the emulsion to guard against spillage and the possible fouling of drains.

There are many different considerations to be taken into account when the removal of swarf from cutting fluid is being contemplated, considerations quite different from those associated with the actual choice of the individual types of equipment. The first and foremost is the degree of clarification required in the cutting fluid following treatment, the second is the number of machines it is desired to service and the third is the capital outlay and the future operating and maintenance costs.

Taking a small gearcutting section as an example, Figure 9.1 shows alternative schemes that embody the three points just mentioned. In scheme number (1) it is very likely that the degree of filtration required will determine whether A should be a filter or a centrifugal separator. If really fine filtration is required then it may be the centrifugal separator that is chosen because of the filter medium 'blinding' too quickly with full throughput. Centrifugal separators of the correct capacity will maintain efficient separation with high flow rates; they are more expensive at initial installation but their running costs are less than those of filters.

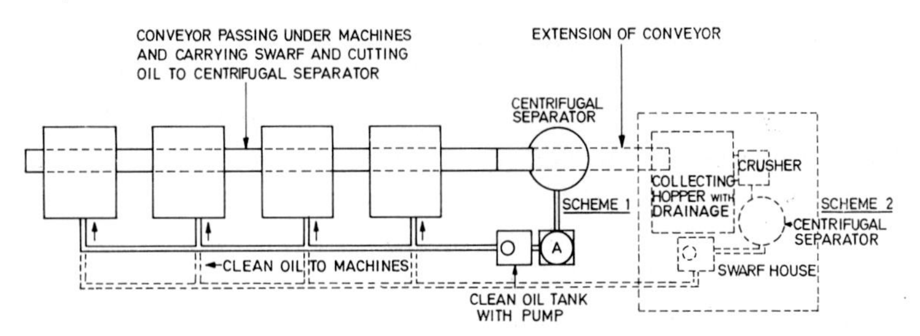

Fig 9.1 Alternative schemes to deal with swarf and cutting fluid in a small gear-cutting section

Still considering scheme number (1) where the separation takes place at the machine, the equipment could take several different forms such as:

1 a low-throughput filter using disposable paper media;

2 a high-efficiency, low-throughput centrifugal separator operating on a by-pass system;

3 a low-efficiency, high-throughput centrifugal separator dealing with the full volume of cutting fluid from the machines.

Although the criterion for selection of equipment might well be the basic capital outlay, experience shows that it is more likely to be the degree of clarification, i.e. the acceptable level of swarf particles remaining in the cutting fluid and being within a given size range.

Moving on to the rather more elaborate scheme number (2), this may be arranged in a simpler manner than the schematic layout suggests, with the swarf house still fully equipped and not occupying too much space. A plan view is given in Figure 9.2.

This arrangement will be capable of handling all the swarf produced in a medium-sized machine shop, cleaning the cutting oil from the swarf chips and removing the fine particles from the fluids as they pass through the centrifugal oil clarifier. Having very much in mind that each ton (1,000 kg) of swarf may contain from 40 to 80 gallons (180–360 litres) of cutting oil valued at approximately six shillings per gallon and that the plant will treat 1 to 2 tons (1,000–2,000 kg) of swarf per hour, then the value of the recovered oil could be in the region of £7,000 annually. From this it will be seen that the capital and installation costs are recoverable in a short space of time, even allowing for plant upkeep and maintenance, and neglecting the higher prices obtainable for the cleaned and dried swarf.

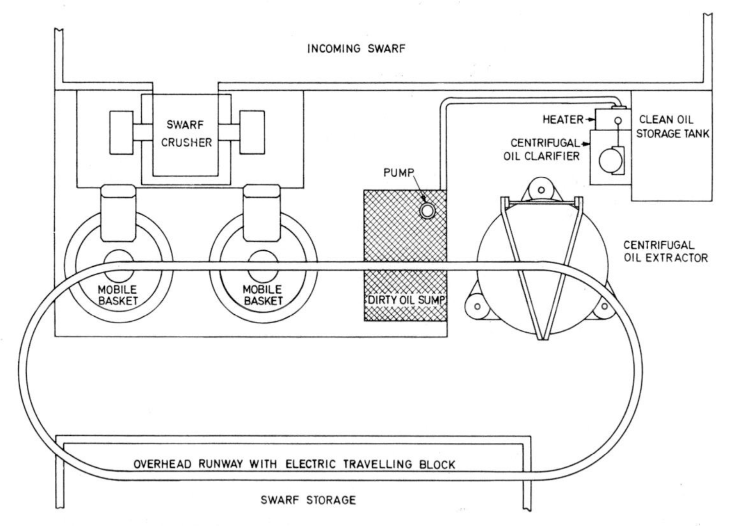

Fig 9.2 Layout for small swarf-processing unit

When the installation of centrifugal separating equipment is being contemplated, it is very necessary that the manufacturer be consulted as it is very difficult to provide performance characteristics such as speed/separation statistics because these are meaningful only for a particular solid-liquid combination. In order that the optimum results may be obtained from centrifuging equipment, samples of the mixtures to be treated should be made available to the manufacturer who will gladly process the mixture through his test plant before putting forward his recommendations.

CENTRIFUGAL SEPARATORS

In order to understand the process of centrifuging, it is necessary to know something of the fundamentals that govern the operation.

A centrifugal separator is a piece of rotating machinery in which is created a high centrifugal force which accelerates the rate of settling-out of solids within the machine. This high centrifugal force within the separator makes possible the separation of dissimilar constituents of a mixture such as solids from liquids or liquids from liquids.

The basic law governing separation by this method is Stokes' Law, this being a physical law which governs the velocity of a particle falling under the action of gravity through a liquid. The particle will accelerate until the frictional drag of the liquid just balances the gravitational force on it, after this it will continue to fall at a constant velocity, known as the terminal or free-settling velocity. This is stated as follows:

$$V = \frac{2r^2(\rho - \rho_1)g}{9h}$$

where V = the settling (or separation) rate
r = radius of particle
ρ = density of particle
ρ_1 = density of the surrounding medium
h = viscosity coefficient of the surrounding medium
g = gravitational acceleration.

In a centrifugal separator the centrifugal force replaces the gravitational force. For any given application, the throughput and the acceleration are the only easily variable factors and it is upon these that a centrifugal separator relies for its operation. By the action of the high centrifugal force the solids are thrown outwards to the inner wall of the centrifuge bowl and, in addition, the heavier liquid, usually water, will separate at a larger radius than the lighter-phase liquid which is usually oil. For efficient operation, it is essential that the phases have different specific gravities and are not miscible.

As the settling rate, V, is expressed in terms of distance/unit time, it follows that the rate of separation can be increased either by reducing the settling distance or by increasing the speed of rotation of the bowl of the centrifuge. A method of reducing the settling distance is to introduce a series of conical discs into the centrifugal separator or, as an alternative, increase the resonance period by producing a longer centrifugal unit of smaller diameter.

Arrangements can be made whereby two separated liquid phases are automatically discharged from the machine, whilst the extracted solids can be either automatically discharged or manually removed at appropriate intervals, depending on the type of unit.

The efficiency of separation depends to a great extent on the oil forming the liquid phase; if a reduction in the viscosity can be achieved, e.g. by increasing the temperature of the oil, then the rate of liquid/solid separation will increase. Experiments have also shown that there is a progressive increase in contaminant removal with increasing quantities of swarf particles in the mixture being treated, and it is considered possible that this may be due to a tendency towards coagulation produced by the electrical charges on the surfaces of the particles.

SELECTING A SUITABLE CENTRIFUGAL SEPARATOR

The general classification of the types of centrifugal separators may be by some particular feature of their construction or by the nature of their solids-discharge pattern. In the former case this is very clearly shown in the detailed descriptions of centrifugal separators given later in this chapter whilst the latter method places the machines into four categories as follows:

1 Batch (basket or peeler type)—the separator must be stopped for manual discharging or have automatic means of doing this. In peeler-type centrifuges the solid cake is removed from the inside of the bowl by plough blades or reciprocating knives.

2 Intermittent—this type incorporates a discharging mechanism that may be manually operated or be self-operating through a series of external valves.

3 Semi-continuous—this type may employ a pusher mechanism or a vibrating bowl to encourage discharge, or a combination of both.

4 Continuous—discharge from this type may be by open nozzle, by open nozzle with heavy-phase recycle or by gravitational methods. In some instances a scroll is used or nutating paring discs or the bowl may be inclined.

The choice of a centrifugal separator should never be made without prior consultation with the manufacturer. However, before the manufacturer is able to advise a prospective user on the type of machine he should install to meet his particular requirements, a number of facts must be established so that the specification can be drawn up, the following information being most necessary:

1 the anticipated flow rate;

2 the types of cutting fluid that will need to be treated;

3 the viscosities of the cutting fluids;

4 the temperature of the mixture at entry to the centrifuge;

5 the types of swarf that will be present in the mixtures;

6 the size and nature of the swarf particles;

7 the anticipated average quantity of swarf that will be present in each mixture;

8 the degree of clarification required;

9 the proportion of water to be removed from the cutting oils, if any;

10 the proportion of soluble oil to be removed from neat cutting oils, if any;

11 the proportion of soluble oil and water to be removed from the neat cutting oils, if any.

Where treatment by centrifugal separator of a swarf and mineral oil mixture is contemplated, it may be necessary to provide additional information on the oil specification as it is quite possible, so stringent is the action of some centrifugal separators, that some of the additives will tend to separate out. This could render the recovered cutting oil incapable of performing the functions for which it was originally designed or, in the case of soluble-oil emulsions, there could be a tendency towards excessive foaming.

If it is considered that a sieve-basket type of centrifugal separator is suitable for a particular application, then care in selection of the operating speed needs to be taken. The highest possible speed does not necessarily provide the maximum separation rate or the optimum washing or drying of the swarf. High centrifugal force can compress a cake of solids too tightly and thus reduce space between individual solid particles and interfere with the drainage of the liquids. Excessive compression can also have the effect of creating fine capillaries between solid particles and this too can prevent the attainment of minimum moisture content. Therefore, instead of thinking only of the maximum centrifugal acceleration, it is necessary to choose carefully the most suitable speed range to suit the mixtures to be treated so as to obtain maximum separating results.

The degree of clarification of the cutting fluid to make it suitable for re-use is important and has economic implications. Centrifugal separators are obtainable that are capable of removing solid contaminants down to the lower micron sizes but the question in many cases is whether such a degree of clarification is necessary, as machines of this standard will require the greatest capital outlay and the running costs will probably be higher. It is contended that a low-to-medium-speed centrifugal separator that operates with a centrifugal force of up to 1,000 g will give clarification of the cutting fluid to a degree that is satisfactory for 80% of the applications in general engineering machine shops. However, where high-class surface finishes are required and have to be produced by fine grinding or honing, then the cutting fluid must be virtually free from solid contaminants and this means using high-speed centrifugal separators capable of producing forces of 5,000 g and upwards.

Although the financial implications of recovering cutting oils from swarf are dealt with in Chapter 10, the figures given in the table on page 210 must be of great interest.

If centrifuging is to be used as the means of removing the bulk of the swarf from the cutting fluid and the fluid from the swarf, as well as clarifying the fluid prior to re-using, then two separate machines may be needed. The first of these centrifugal separators would be used to separate the coarse swarf from the cutting fluid and might well be of the basket type, the degree of contaminant removal being down to 10–20 microns approximately. The discharge of the extracted solids could be either by manual or automatic means.

TABLE 9.1 *Cutting-oil recovery in gallons per hour assuming a retained-oil content of 3% on the swarf after centrifuging.*

Cutting-fluid content of swarf—% by weight	*Rate of feed—lb per hour*							
	4000	*3500*	*3000*	*2500*	*2000*	*1500*	*1000*	*500*
30	160	140	120	100	80	60	40	20
28	148	129	110	92	74	55	36	18
26	136	118	100	85	68	51	34	17
24	124	107	90	77	62	46	31	16
22	112	96	84	70	56	42	28	14
20	100	85	75	63	50	37	25	13
18	88	74	66	55	44	33	22	11
16	76	63	58	48	38	29	19	10
14	64	52	49	40	32	24	16	8
12	52	41	38	33	26	20	13	7
10	40	30	28	25	20	15	10	5

If it is necessary for the cutting fluid to have a greater degree of cleanliness then a second operation will be called for. For this purpose a different type of centrifugal separator will be needed and this could be one of the following types—disc bowl with manual cleaning, self-cleaning disc bowl, or multi-chamber bowl. These types will accommodate throughput rates from 35–90 gallons per minute and will remove solids down to 1–2 microns from soluble-oil emulsions. In the case of oils having a higher viscosity the minimum particle size which can be separated from the oil is approximately as follows: 2 micron with 50 seconds Redwood 1; 4 micron with 250 seconds Redwood 1; and 6 micron with 500 seconds Redwood 1.

TYPES OF CENTRIFUGAL SEPARATOR

Figures 9.3, 9.4, 9.5, 9.6 and 9.7 show five different types of centrifugal separator.

1 *Basket type*

The basket-type centrifugal separator is a liquid/solid separator having a simple rotating bowl giving a large solids-holding capacity. The bowl speed is usually in the range 1,450–2,900 r.p.m. which, combined with the large solids-holding capacity, makes this type of

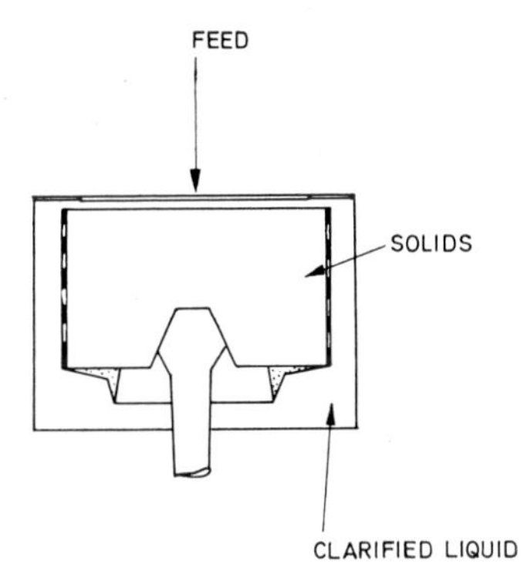

Fig 9.3 Basket-type centrifugal separator

separator well suited to the treatment of soluble and light mineral oil cutting fluids, although giving a lesser degree of separation than would be obtainable with the other centrifugal separators described.

Individual designs incorporate easily removable sludge containers or automatic solids discharge.

2 *Tubular-bowl type*

The tubular-bowl centrifugal separator is suitable for liquid/liquid or liquid/solid separation and consists of a tall, comparatively

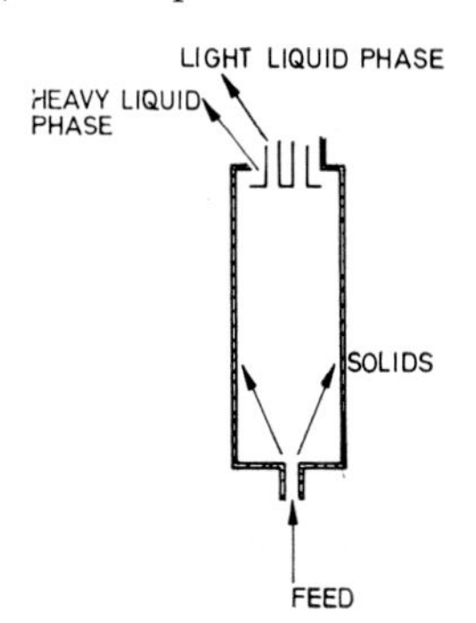

Fig 9.4 Tubular-bowl-type centrifugal separator

small-diameter bowl rotating at high speed. Liquid feed is generally from the bottom with liquid discharge from the top, the solid being retained in the tube. The length of the tube ensures sufficient holding time to achieve the required degree of separation.

3 *Disc-bowl type*

These machines, with separating compartments (bowls) running at speeds of 4,500–6,500 r.p.m. and generating centrifugal forces of 5,000 g and upwards, separate instantly mixtures of two liquids or liquid(s) and solids very efficiently.

The bowl consists of a body and hood held together by a lock ring. Mounted coaxially inside the bowl is a hollow distributor into which the liquid to be separated is fed. Outside the distributor is a stack of conical intermediate discs through which the feed liquid

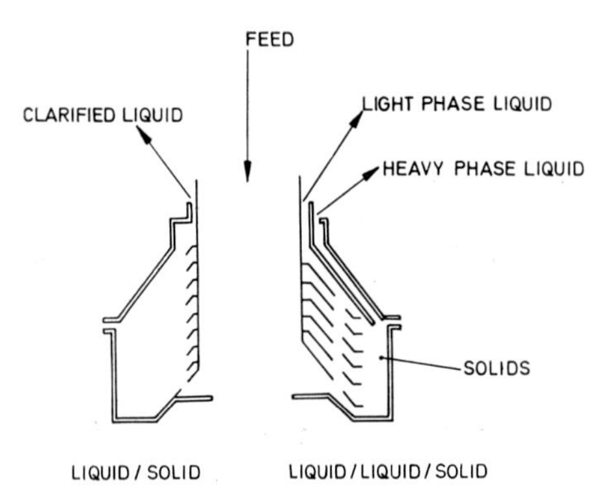

Fig 9.5 Disc-bowl type centrifugal separator

rises and where the actual separation takes place. These have the effect of dividing the separation zone into a series of very thin layers, which means that the different phases have only a very short distance to travel in order to free themselves from each other. The light-liquid phase moves inwards and passes up to the outlet at the neck of the bowl, while the heavy liquid and/or solids move outwards into the space outside the disc stack. In liquid/liquid separation, the heavy liquid flows round the edge of a special top disc and passes between this disc and the bowl hood to a second outlet arranged concentrically around the first, while solids are thrown into the sludge space at the periphery of the bowl and retained.

4 *Self-cleaning disc-bowl type*

The self-cleaning separator has gained wide acceptance in recent years, thanks largely to its labour-saving features. Chief among these is the special design of the bowl, which can be opened at intervals to discharge accumulated solids while still running at full speed; in addition, the discharge cycle can be controlled from an external timer, and the machine can be cleaned in place by circulation of cleaning liquids. All this adds up to long running times and few shut-downs for maintenance. Feed and discharge of liquids are continuous.

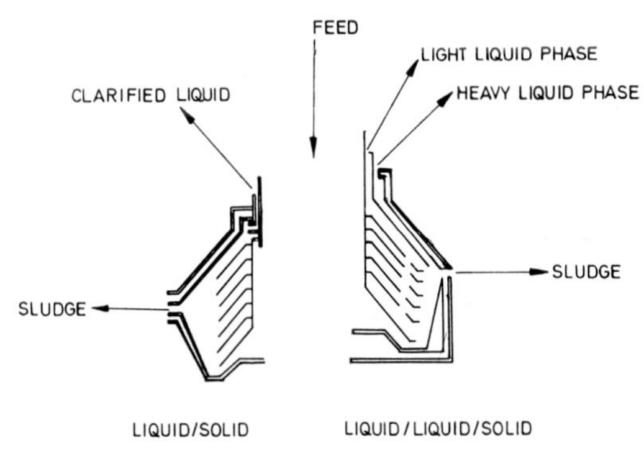

Fig 9.6 Self-cleaning type centrifugal separator

The basic constructional features comprise a vertical disc-insert bowl of double conical shape with an annular discharge opening at the widest part. A sliding bottom or sealing-ring which in the operating position is held closed by the hydraulic pressure of an operating liquid (usually water) is situated in the compartment underneath. When lowered, an annular discharge opening is formed between the sliding bowl bottom and the bowl hood, allowing all or part of the contents of the bowl to escape. The bowl is surrounded by an annular spiral or deep collecting cover from which ejected material is discharged by gravity.

Solids accumulate in the sludge space at the junction of the sliding bowl bottom and the bowl hood. At suitable intervals operation of the hydraulic circuit permits the bowl bottom or ring to drop rapidly and the bowl contents are ejected by centrifugal force. Depending on the length of time the bowl remains open, either total discharge (of the entire contents of the bowl) or partial discharge (solid phase only) may be obtained. In the former case the feed is temporarily interrupted during the discharge cycle and the bowl can also be flushed with water, which escapes with the solids; in the latter case, separation continues uninterrupted throughout.

5 *Multi-chamber type*

The multi-chamber centrifugal separator is designed for liquid/solid separation, and in general appearance and construction is similar to the disc-bowl model but the disc pack is replaced by a series of concentric, vertical cylinders with alternate top and bottom fixing to form chambers through which the liquid flows from the central feed point to the outer diameter.

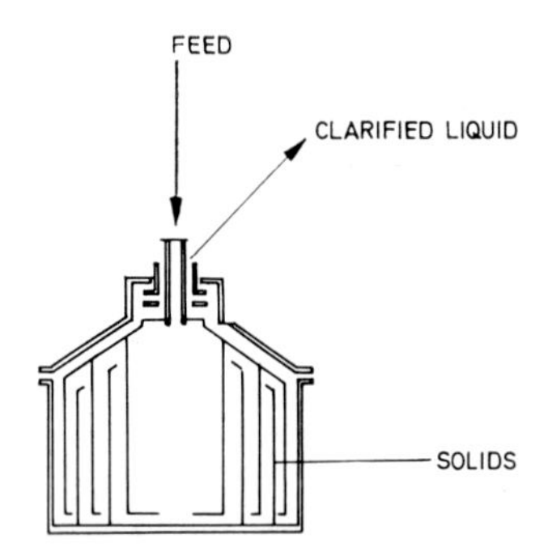

Fig 9.7 Multi-chamber type centrifugal separator

The multi-chamber separator is designed to have a large solids-holding space for a given diameter bowl. The degree of separation which may be achieved is lower than with a disc bowl or tubular separator but it has the advantage that less-dense particles, which would otherwise be carried through the centrifugal separator, are retained against the outer diameter of the chamber walls.

HYDROCYCLONES

Hydrocyclones are designed for liquid/solid separation when the liquid has a low viscosity.

The separating effect in a hydrocyclone is dependent upon centrifugal force as in the centrifugal separator but, in this case, it is not generated by mechanical means. Also, as with the centrifugal separator, the separating effect is governed by Stokes' Law. The design of a hydrocyclone takes the form of an inverted cone which has an entry for the feed positioned tangentially to the major diameter, and centrally placed openings at the top and bottom for the ejection of the separated constituents—see Figure 9.8. The centrifugal force is actually created when the liquid mixture is

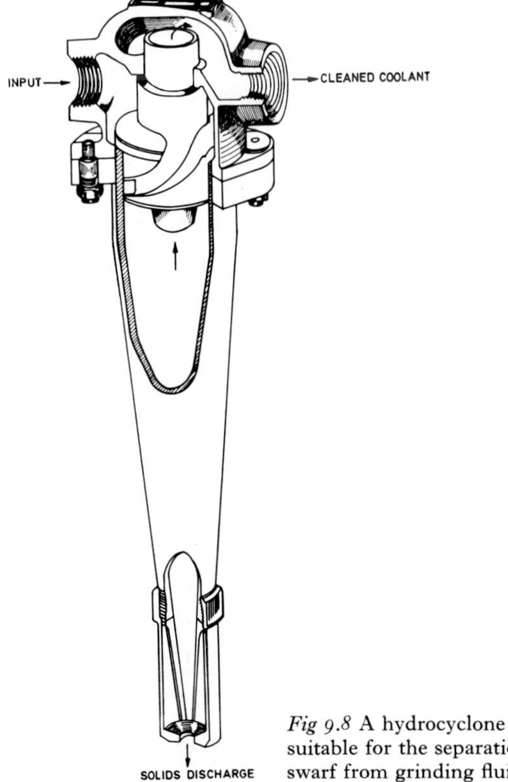

Fig 9.8 A hydrocyclone suitable for the separation of swarf from grinding fluids

injected tangentially through the feed opening into the tapered cyclone chamber, a gravitational acceleration is set up and the angular velocity at the tapered walls produces a hyperbolic increase in the velocity. Because of the tapered form of the cyclone chamber and the centrifugal force, a combination that creates a vortex, the solid swarf particles are thrown against the chamber walls causing them to lose their velocity and to slide downwards to the discharge opening. At this stage of the process the clarified liquid has also reached the bottom of the cyclone chamber where, instead of being discharged with the solids, it is guided into the formation of an inner vortex which carries it upwards through the centre of the chamber for discharge through the outlet at the top. The contaminants discharged from a hydrocyclone are not dry as with a centrifugal separator, but are in the form of a liquid with a high concentration of solids.

Although this type of separation equipment is capable of operating at around 90% efficiency when dealing with swarf and cutting fluids, giving clarification down to 10 micron with soluble-oil emulsions and cutting oils of low viscosity, only now is it achieving any real popularity when compared with centrifuges and filters. In Europe generally, there is an ever-increasing awareness of the possibilities attaching to the use of hydrocyclones for the cleaning of cutting fluids and many models are now on the market.

For machine-tool applications, hydrocyclones are available in convenient sizes that are comparatively cheap and are simple in operation. They are suitable for application as single units or in pairs and can be used as multiple 'banks' to cater for the cutting fluids from a number of machines. The advantages to be gained from the use of hydrocyclones are enumerated as follows:

1 There are no moving parts to wear out or break down.

2 Where the abrasive properties of the swarf particles could cause wear, such sections can be constructed from heavier plate materials or castings or indeed from ceramic materials or ceramic coatings.

3 It is possible for known areas of heavy wear to be designed and made as replaceable sections.

4 Relatively high temperatures in the swarf/liquid mixture at entry to the hydrocyclone do not affect the operation.

5 Efficiency for a given set of conditions is consistent and, since there is nothing to go wrong or vary mechanically, it will be maintained as long as required.

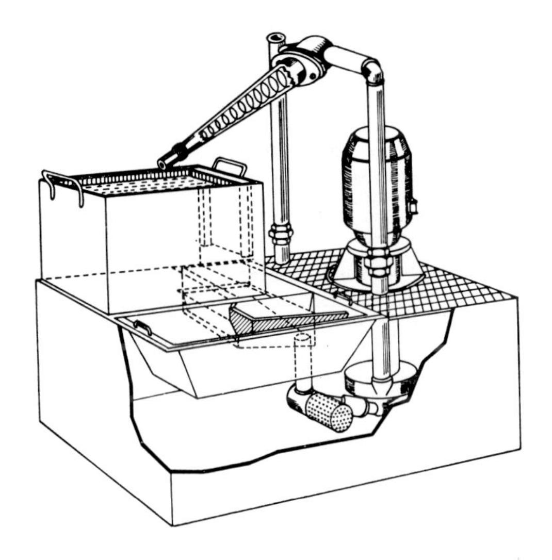

Fig 9.9 A method of using a hydrocyclone for continuous cleaning of the cutting fluid

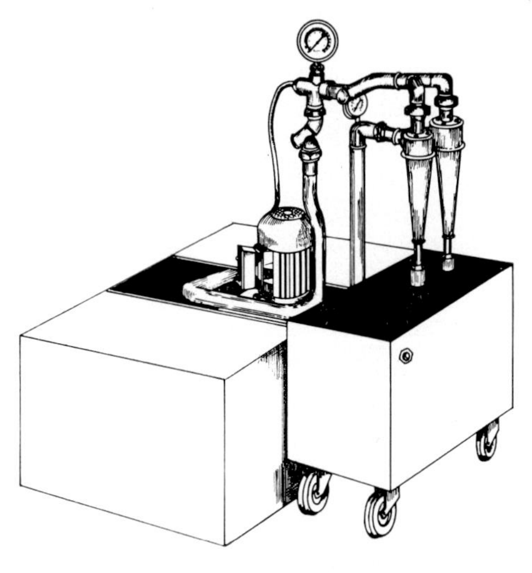

Fig 9.10 A pair of hydrocyclones arranged for the continuous cleaning of cutting fluids

6 Within limits, an increase in the concentration at the inlet will not lower efficiency.

7 Maintenance costs are almost non-existent.

Figures 9.9 and 9.10 show two methods of using hydrocyclones for continuous cleaning of the cutting fluid.

PERFORMANCE

Particulars of the characteristics and performance of the different types of centrifugal separators are shown in Table 9.2.

TABLE 9.2 *Summary of characteristics of different types of centrifugal separator*

Type	*Suitable for separation of*		*Through-put range g.p.m. (l min⁻¹)*	*Max. % solids in feed depending on through-put*	*Solids removal in microns*				*Remarks*
	*L/S**	*L/L/S***			*Soluble-oil emulsion*	*50 secs Redwood I*	*250 secs Redwood I*	*500 secs Redwood I*	
Hydro-cyclone	√		20–2000 (90–9000)		10	10	Not suitable	Not suitable	Gives 98% extraction of solids down to 10 micron
Basket	√		Up to 60 (270)		5	5	Not suitable	Not suitable	Self dis-charge of solids
Basket	√		2–40 (9–180)		5	5	8	Not suitable	Self dis-charge of solids
Tubular bowl	√	√	Up to 16 (75)	1–2	1–2	2	4	5	Low main-tenance costs due to mech-anical sim-plicity
Disc bowl	√	√	Up to 60 (270)	1	1–2	2	4	6	Manually cleaned machines
Disc bowl	√	√	Up to 36 (164)	1	1–2	2	4	6	Manually cleaned machines
Disc bowl	√	√	Up to 70 (318)	1	1–2	2	4	6	Manually cleaned machines

TABLE 9.2 *(continued)*

Type	*Suitable for separation of*		*Through-put range g.p.m. (l min⁻¹)*	*Max. % solids in feed depending on through-put*	*Solids removal in microns*				*Remarks*
	*L/S**	*L/L/S***			*Soluble-oil emulsion*	*50 secs Redwood I*	*250 secs Redwood I*	*500 secs Redwood I*	
Self-cleaning disc bowl	√	√	Up to 90 (400)	5	1–2	2	4	6	Intermittent solids discharge
	√	√	Up to 38 (170)	6	1–2	2		6	Continuous solids discharge
	√	√	Up to 32 (145)	1	1–2	2	4	6	Intermittent solids discharge
	√		Up to 90 (400)	5	1–2	2	4	6	Intermittent solids discharge
		√	Up to 70 (318)	5	1–2	2	4	6	Intermittent solids discharge
Multi-chamber bowl	√		Up to 35 (160)	1	1–2	2	4	6	Between 2 and 6 chambers
	√		Up to 45 (200)	1	1–2	2	4	6	

* L/S = Liquid/solid
** L/L/S = Liquid/liquid/solid

10 THE PROCESSING AND HANDLING OF SWARF

INTRODUCTION

In many thousands of small engineering plants, swarf produced in the machine shop is manually removed and dumped at a convenient place in the works yard from whence, in due course, it is collected by a scrap-metal merchant. At the other end of the scale, there are very large machine shops which have installed complete and specialised swarf-handling facilities to deal with all types of swarf in the large quantities that are produced; such installations may be capable of crushing the swarf, passing it through centrifugal separators to remove oil and cutting fluid, delivering it to storage bunkers to await collection, and may be almost fully automated.

However, in between the limits outlined above there are the engineering concerns that are of medium size to whom the removal and disposal of swarf is still very much a problem embodying an economic aspect. It is hoped that the information contained in this section will assist in the selection of suitable equipment for a medium-sized machine shop and give an indication of the financial savings, together with other benefits, that can accrue from the use of properly designed and chosen swarf-handling equipment. However, because of the diversity of products and the materials from which they are made, it must be realised that there is no one standard plan that will suit all machine shops; the contents of this chapter will act as a guide only and emphasis must be placed on the need to discuss the problem fully with the equipment manufacturers.

THE ECONOMIC ASPECTS OF SWARF PROCESSING

When considering the implications of installing a swarf-processing facility, the prospective user must give serious consideration to the capital cost of the equipment that would be needed and the time required to recover his outlay. It may well be that knowledge of the financial benefits that can be gained from processing swarf is not generally appreciated and the following may assist in clarifying the position:

1 Swarf that is produced with mineral oil as the cutting fluid may contain from 40 to 80 gallons (180–360 litres) of oil per ton (1,000 kg), the majority of this quantity being recoverable. An efficient centrifugal separator is capable or removing from 95–97% of this oil, the average amount normally remaining on the swarf being in the region of 3%.

If it is assumed that adequate processing by a centrifugal separator will remove at least 40 gallons from 1 ton of swarf, and that the value of oil is from 5 to 6 shillings per gallon, then the savings will be from 200 to 240 shillings per ton of swarf processed. In calculating the potential saving the user should take into account the fact that 95% of the oil used will be recovered, the remaining 5% being lost in unaccountable directions, i.e. workpiece, workshop floor, oily rags, etc.

Depending upon the degree of extraction achieved by the separating equipment, the savings could well be greater than in the example given above.

2 The use of cutting tools equipped with chipbreakers is the way in which to avoid producing bushy swarf, this being inconvenient to handle, difficult to process and not easy to sell. However, if bushy swarf is produced then it may be 'crushed', which converts the long helices of swarf into small chips and segments that are easier to handle and which take up far less space—see Figure 10.1. There is no difficulty in disposing of swarf in this condition and metal merchants are usually prepared to buy at a premium price of 35 to 50 shillings per ton (1,000 kg) extra for swarf in this form.

Crushed swarf requires less space and this means that a saving in transport costs may be made, whether it be by road or by rail.

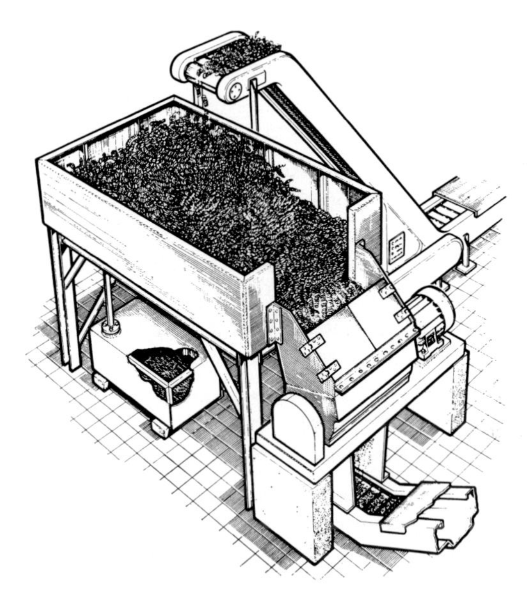

Fig 10.1 Small swarf-processing unit incorporating a crusher and conveyors

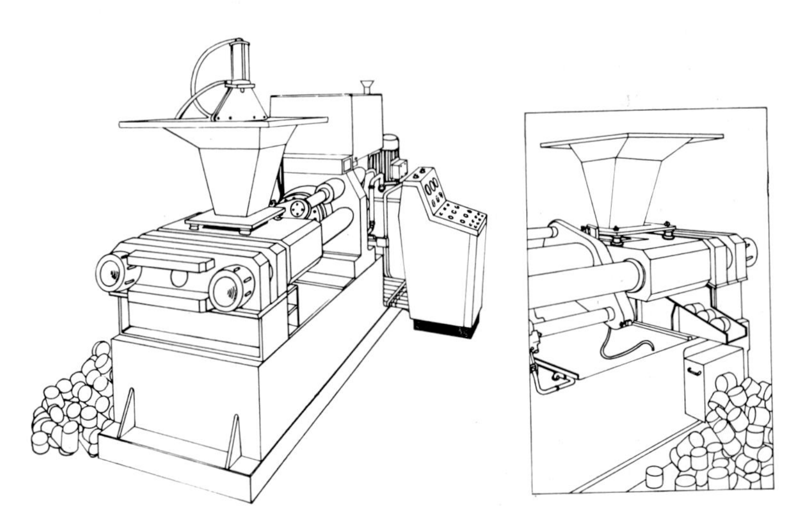

Fig 10.2 A type of briquetting machine

It is quite possible for savings of 15 to 25 shillings per ton (1,000 kg) to be made, depending on the distances involved.

3 Another form of swarf processing involves compressing the material into briquettes—see Figure 10.2. There is a growing demand for swarf in this form for re-melting in electric arc furnaces and, again, a somewhat better price should be obtainable.

One of the difficulties to be encountered here is that the purchasers of such briquettes may require the residual oil content to be as low as 0·05% and this figure cannot be readily achieved using conventional centrifuges. One method of attaining this low oil-content figure is to wash the swarf in a suitable detergent liquid and to dry it in an airstream.

When planning a swarf-processing facility of substantial size, it is recommended that enquiries be made concerning the possibilities of swarf briquetting and any financial benefits accruing.

4 With swarf processing there are many intangible benefits that are not easily detailed but which must be taken into account. The current high cost of non-productive labour and its scarcity is a problem that faces all managements and there is little doubt that a more effective method of handling and processing swarf will lead to measurable economies in this direction. Less swarf and cutting fluid around the machines also means less labouring time required for cleaning and provides for better housekeeping—see Figure 10.3.

Better housekeeping means cleaner floors and this reduces the hazards of slipping by personnel or skidding by vehicles.

CONSIDERATIONS PRIOR TO SELECTION OF EQUIPMENT

When capital is to be expended on the purchase of swarf-processing equipment, it is important that the best value for money be obtained after taking every possibility into consideration. This being so, the following points must be borne in mind:

1 Every effort should be made to avoid the production of bushy swarf as, in this form, it is difficult to handle, cannot be separated from the cutting fluid, cannot be easily disposed of and does not command a good price from the metal merchant.

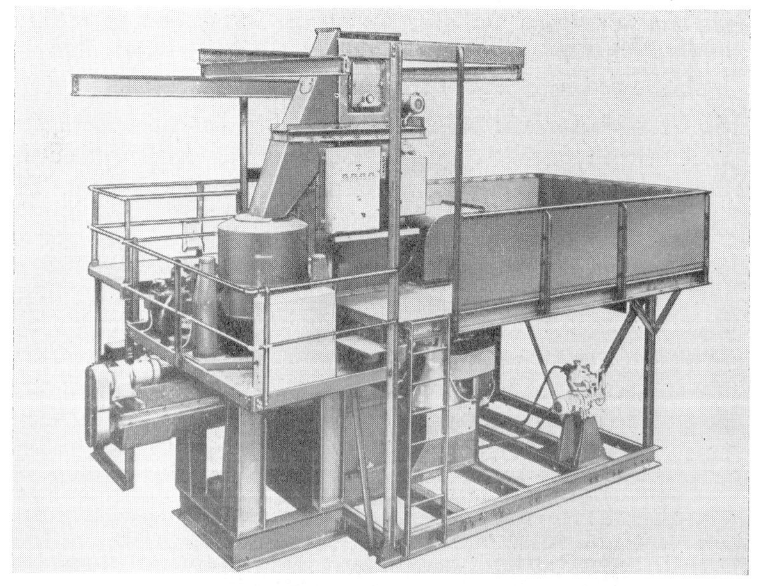

Fig 10.3 A swarf-processing installation with conveyors, hoppers, crusher, centrifugal separator, and drainage lines for cutting fluid

2 If bushy swarf must be produced, then crushing or breaking is recommended as an aid to further processing. This needs consideration, however, because it may not be economic to install such equipment unless the quantity of swarf is in the region of at least one ton per hour (this is assuming that one man will be engaged full time in attendance on the crusher). However, in a plant of medium size, where swarf is likely to be manually removed from the machine shop to the site of the crusher, it is often arranged that the swarf is allowed to accumulate on the site during the day and one of the labourers bringing out the swarf will run the crusher for perhaps two separate periods of one hour or so, thus disposing of the accumulation. By this method, a crusher can be an economical proposition when required to process less than one ton of swarf per hour and the capital cost may be quickly recovered from the enhanced selling price of the swarf.

3 The cutting fluids recovered from swarf by centrifuging will usually require a further cleaning operation by either liquid

clarifiers or filters and this means the installation of ancillary equipment. Where neat cutting oils are recovered they will benefit by being pasteurised after the final cleaning operation.

4 Because of recent legislation controlling the disposal of industrial wastes, referred to in detail in Chapter 11, water-based cutting fluids will need treatment and separation before disposal. This again will call for extra equipment.

5 When planning the layout of a swarf-processing unit and bearing in mind that space in any plant is usually at a premium, thought should be given to the possibility of locating liquid tanks, pumps, filters, etc., around or even under the crushing, centrifuging and associated handling equipment. By doing this it is often possible to reduce the total floor area required. Figure 10.4 shows an elaborate swarf and cutting-fluid installation.

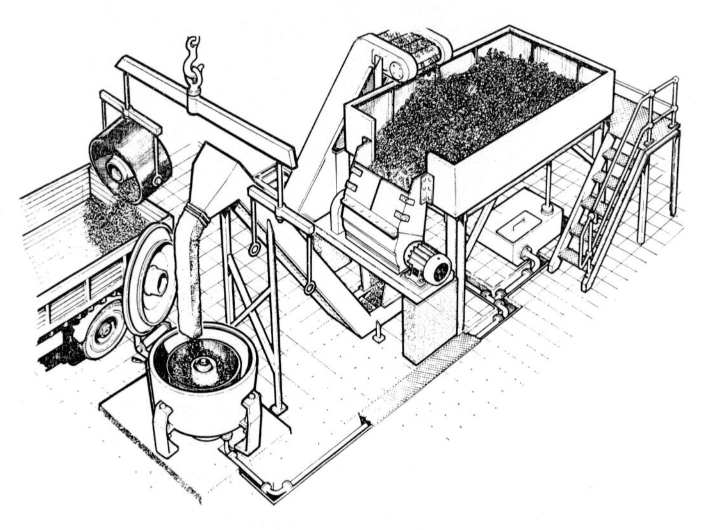

Fig 10.4 An elaborate swarf and cutting-fluid processing installation

6 It may be that the planned processing unit will be called upon to deal with a variety of cutting fluids with the attendant dangers of cross contamination, and this implies that at least as much care must be exercised in the selection of the liquid-handling equipment as in the choice of the swarf-handling equipment. It is

desirable, therefore, to scrutinise the current cutting-fluid utilisation in relation to any possible difficulties that may be encountered when the processing unit is put into operation. The advice of the cutting-fluid suppliers should be sought at a very early stage in the investigation and due note should be taken of the need to recover and re-use the maximum amount of cutting fluid.

EQUIPMENT COMPRISING A SWARF-PROCESSING UNIT

The amount of equipment used to make up the facility will obviously depend upon the size of the manufacturing plant and the degree of processing that is deemed necessary. Practically all facilities will incorporate conveyors and elevators, centrifugal separators and filters, and ancillary equipment such as pumps, and this machinery is dealt with in further detail in previous chapters of this manual. Figures 10.3 and 10.4 are of two typical swarf-processing installations, the layout having been expanded to allow easier identification of the individual pieces of equipment.

However, mention has been made of other pieces of equipment and a few brief comments may be helpful, together with the illustrations.

1 Bushy swarf may be readily reduced to manageable chips and particles by passing it through a hammer mill or crusher. The hammer mill is basically a hopper lined with hard-wearing material against which the swarf is pulverised by the beating action of the 'hammers' attached to a shaft which revolves at high speed within the hopper. A drawback to this type of machine is the ease with which the beaters break if the swarf contains any bar ends or other large pieces of solid metal.

The action of crushers is very different and the name 'crusher' aptly describes this. The inside of the crushing chamber is usually equipped with cutters which are bolted to the walls at appropriate angles, a shaft carrying further cutting arms revolves inside the chamber and the cutters rip and tear the long helices of swarf into small pieces. These machines are usually equipped with slipping clutches which prevent internal damage should the cutters become locked by bar ends, bolts, etc., contained in the swarf; some are arranged so that when a blockage occurs the direction of rotation is automatically reversed to allow the solid object to be released. Crushers may be horizontal or vertical and most are arranged for either manual or mechanical feeding—see Figures 10.5 and 10.6.

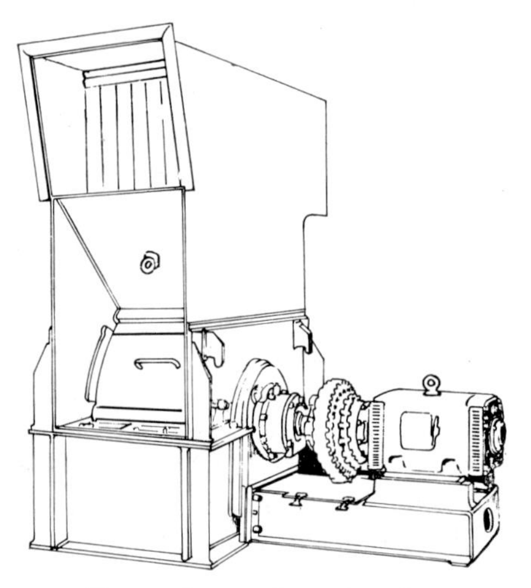

Fig 10.5 A type of swarf crusher

2 Hoppers for containing the swarf at various stages of the processing play an important part and close consideration should be given to their design.

The hoppers into which the swarf may be fed immediately on its arrival from the machine shops must be liquid-tight in order to contain the cutting fluids that will be adhering to the swarf. If the swarf is in bushy form, it will normally feed from this first hopper into the crushing equipment; both the hopper and the crusher should have suitable arrangements to take away the cutting fluid that will drain from the swarf.

If the broken swarf is to be held in a secondary hopper to await centrifuging, then this hopper must also be made liquid-tight and have drainage arrangements.

It is probable that, following the centrifuging operation, the cleaned swarf will need to be stored until a truck load has accumulated. A convenient way in which to hold the swarf is in an overhead hopper so located that direct discharge into lorries or rail trucks can be made. After centrifuging, the swarf should be comparatively dry and free from cutting fluids but it is recommended that this hopper should be suitably drained. The discharge doors should be designed for easy manipulation; swarf in bulk can become locked together by mechanical action and will not slide freely; so large sliding bottom doors or clam-shell outlets are the answer.

3 The possibility of financial gain from the sale of briquetted swarf has already been mentioned and some notes on the type of equipment obtainable may prove useful. It is possible to make briquettes from the swarf produced during the machining of most metals and also to make bales from the off-cuts and clippings resulting from sheet-metal presswork. Typical briquetting machines are shown in Figures 10.7 and 10.8.

Fig 10.6 Another type of swarf crusher

The machines used for this operation vary widely in both type and size but machine-shop swarf is usually compressed into 6 in (150 mm) cubes or into cylindrical briquettes of about 4 in (100 mm) diameter and up to 6 in (150 mm) long. The method of operation is simple. The swarf is fed manually or automatically into a hopper that is situated above the compressing chamber. When the chamber is filled with the untreated swarf it is closed and the internal horizontal and vertical dimensions control one or two sizes of the ultimate briquette. Pressure is applied within the chamber by a ram which may be hydraulically or mechanically operated and outputs of up to 500 briquettes per hour are quite possible.

If these machines are to form an integral part of a processing facility then they will be permanently sited, but they are also obtainable in portable form.

Fig 10.7 A vertical type of briquetting press

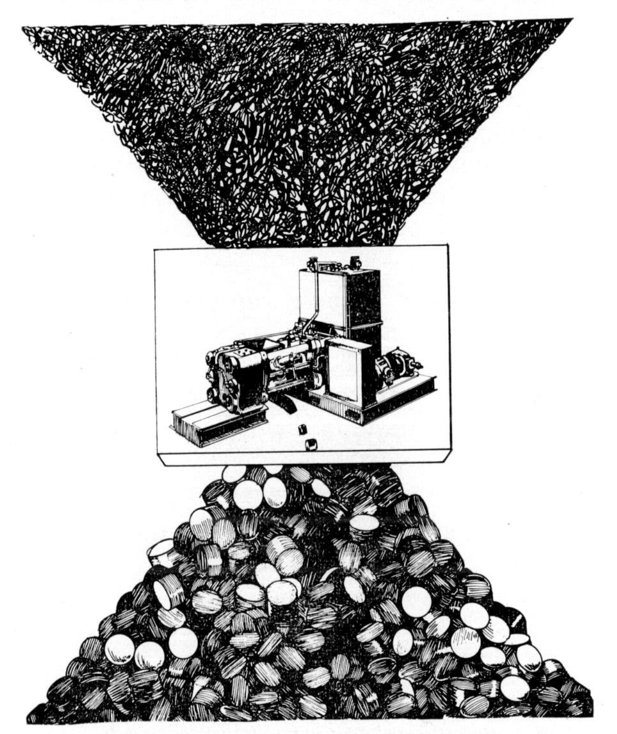

Fig 10.8 Another type of briquetting press

CAPACITIES OF SWARF-PROCESSING INSTALLATIONS

The following suggestions and recommendations concern the type of equipment and size of unit that may be installed to deal with varying rates of swarf production. The information should be treated with a certain amount of reserve since the circumstances can vary so very widely from one factory to another; the intent is to indicate to a prospective user the type of installation he should be considering on the basis of his current or projected swarf-output rate.

Dealing firstly with ferrous materials, the following may apply:

1 Where the output of swarf is less than 8 tons (8,000 kg) per week, it is not recommended that any elaborate processing arrange-

ments be made. Such a volume of swarf should be accumulated in liquid-tight containers and, when full, should be taken away by the scrap-metal merchant. By using such containers better housekeeping is assured, the disposal of the swarf is speedier, contamination of the drains is minimised and dangers to health are reduced. If the proportion of bushy swarf is considerable, it may be advantageous to install a hammer mill or crusher of medium size.

2 In plants where the production of swarf is in the range of 8 tons (8,000 kg) per week or more, and particularly when neat cutting oils are used, the installation of centrifugal separators and oil-purification equipment will usually be economically advantageous, but even at this rate of swarf production elaborate units may not be necessary. Again, in the event of a large volume of bushy swarf being produced, the provision of a crusher could be an asset.

3 In the bigger plants where swarf production can exceed 30 tons (30,000 kg) per week, a fully automated system should be employed, complete with conveyors and elevators, crushers, hoppers, centrifugal separators, filters and a full cutting-fluid reclamation arrangement.

4 If water-based cutting fluids are generally used throughout the plant and swarf outputs are in excess of 20 tons (20,000 kg) per hour, then it could be economic to provide centrifuging capacity along with a suitable fluid-disposal plant. The drying action of the centrifugal separator may result in a better price from the metal merchant.

The situation with regard to non-ferrous swarf is somewhat different and merits attention being given to the following:

1 In considering the degree of processing to be given to this type of swarf, it may again be uneconomic to install a complete unit when the volume of swarf is 3 tons (3,000 kg) per week or less. This may not be the situation when metals such as titanium and other highly priced alloys are being machined; such circumstances call for special consideration and precautions.

2 Even with smaller quantities of swarf that have been produced

with the aid of neat cutting oil, a small, basket-type centrifugal separator is an economic proposition. Such equipment can be purchased for as little as £350 and has a capacity of 30 cwt (1,500 kg) of swarf per hour; brass or gunmetal swarf can contain as much as 8–10 gallons of oil per ton.

3 The swarf produced by turning rolled or extruded aluminium and certain of its alloys is quite likely to be of bushy form and can be a nuisance if produced in any volume, remembering that the swarf produced in chip form from milling 1 cubic inch of aluminium can occupy a volume of 50 cubic inches. Because of its ductility, it is not possible to crush bushy aluminium swarf easily and metal merchants are unlikely to accept it in baled form.

4 The machining of some non-ferrous materials is aided by the use of a water-based cutting fluid. In these conditions the introduction of a centrifugal separator could be considered worthwhile when the amount of swarf is in the region of 20 tons (20,000 kg) per week.

CONCLUSION

In these days of constantly rising costs, it is necessary to explore every possible avenue that may lead to financial economies and no avenue is more worthy of exploration than the processing and disposal of swarf. With the very regular progress being made in the automation of machine tools and the ability of modern cutting tools to remove more metal in shorter times, the problem of swarf removal is likely to grow rather than decline. The situation, therefore, calls for far more consideration today than it has ever received in the past except in the very largest of plants, and correct and efficient processing of swarf produces benefits never previously realised.

11 CUTTING FLUIDS, EFFLUENTS, WATER POLLUTION AND THE LAW

INTRODUCTION

Legislation governing the composition of industrial effluents is beginning to be enforced and such enforcement may create difficulties for the users of cutting fluids. During use, all cutting fluids become degraded to a greater or lesser extent and this leads to their rejection as lubricants, dissipaters of heat and cutting aids. River Boards and Local Authorities are becoming increasingly reluctant to accept wastes containing the fluids used in metal-working operations because of their undesirable properties, and this is the reason for their enforcement of recent legislation with increased severity. Thus, pollution problems arising from the use of these fluids are beginning to influence the choice of fluids and will, in turn, influence machine-tool design and operating techniques.

This chapter is intended to clarify the position about recent legislation, to discuss the situation as regards pollution and effluents, and to endeavour to offer advice on some of the problems that will arise.

POLLUTION AND THE LAW

Pollution is legally established to be the introduction of anything to water which will adversely affect the natural quality of the water in a river or stream. In this respect, oily wastes are particularly undesirable in effluents as the oil floats to the surface and prevents natural aeration. Additionally, and even though the oil be emulsified, it could interfere with biological sewage-treatment processes to an extent that has become severe in the newer types of activated-sludge plants. These dangers must be avoided and the need for this will be more readily appreciated when it is realised that during dry weather a significant proportion of the water in rivers consists of sewage effluent and that these waters will have to be used to an increasing extent as a source of supply for public usage.

As long ago as 1876, legislation produced the Pollution of Rivers Act but little notice was taken of the regulations laid down and the effects were negligible. Of the later Acts, the most significant are the Rivers (Prevention of Pollution) Acts of 1951 and 1961, and the Water Resources Act of 1963. By these Acts, the consent of the appropriate River Authority must be obtained to cover all discharges to surface water in England and Wales. In Scotland the River Purification Boards have similar powers. These Authorities are also given powers over all abstraction of ground water and discharge underground; moreover, riparian owners have certain rights under Common Law and these need to be heeded.

Local Authorities must also comply with the conditions imposed by the River Authorities in respect of the discharge of sewage effluents and, in order to safeguard their interests, under the Public Health (Drainage of Trade Premises) Act, 1937, and the Public Health Act, 1961, they are given powers to control discharges of trade wastes to their sewers. Additionally, some Local Authorities are given powers of control under local Acts. The consent of the Local Authority is required for all discharges of trade effluent, and payment for the collection and treatment may be imposed. Such charges are generally based upon both the volume of waste and the degree of contamination. It is logical, therefore, to keep the volume to a minimum and the quantity of contaminant as small as possible. When payment has to be made for the reception and treatment of industrial effluents, the charges are usually based on the Biochemical Oxygen Demand (B.O.D.) which is a measure of the polluting power of the effluent and, because of the presence of oils and additives, the assessment of such charges is complex.

The legislation referred to does not specify the permissible concentrations of contamination but the limits are usually fixed at 5–10 mg l^{-1} oil, 30 mg l^{-1} solids and B.O.D. at 20 mg l^{-1} for discharge into rivers; for discharge into sewers the concentrations are usually 25–30 mg l^{-1} oil, 100–400 mg l^{-1} solids and 500 mg l^{-1} B.O.D.

Industry must now realise that water is a most important raw material and that careful husbanding of water resources and purification of polluted waters is its responsibility as well as being in its own interests. Therefore, industry must accept that the cost of reducing pollution must be regarded as an integral part of production costs.

THE EFFECTS OF CONTAMINATION BY CUTTING FLUIDS

Most cutting fluids will have as constituents mineral oils, emulsifying agents (which may be phenolic in character), wetting agents, anti-foam agents, rust inhibitors, water conditioners and bactericides. Of these, the oil is the constituent most likely to cause difficulty when the waste waters are insufficiently treated before discharge to a sewer or a river, but the concentration of the phenolic substances is also very important.

If ground water becomes polluted by oils, objectionable tastes and odours may persist for a very long time and it could result in withdrawing from service any wells that become affected in this way. It is, therefore, dangerous to dispose of oily wastes to soakaways or to deposit them on tips where there is the slightest risk of causing pollution to underground waters.

Oil on rivers and streams is aesthetically objectionable and leads to unsightly fouling of vegetation and banks. It is possible that some oils may be harmful to vegetation and so oil-polluted waters are unsuitable for crop spraying or irrigation. Oily waters are a menace to fish and possibly to other animal life. Certain oils and water-soluble additives from them are directly toxic to fish or may taint their flesh, whilst all oils may coat their gills and other epithelial surfaces. In high concentrations, phenolic substances are toxic to fish and other life.

Should effluents containing oil reach a sewage-treatment plant, then serious trouble could result. After receipt at a plant, the sewage undergoes a number of different treatments such as settling-out, biological oxidation, filtering and micro-straining. The residue in the form of sludge is often treated by anaerobic digestion but if oil is adsorbed on the sludge and accumulates in the digesters, then the sludge becomes difficult to dry-out and is unsuitable for agricultural purposes.

So far, only the effects of oil pollution have been considered, but attention must be given to the biological degradation of the cutting fluid and control of the bacteria, particularly the sulphate-reducing variety. Their diet is compounds containing oxygen and sulphur and, unless precautions are taken, they will quickly build up in emulsified-cutting-oil circuits. They break down emulsifiers and corrode machine components; this corrosion can be severe when septic conditions (anaerobic fermentation) develop in pipe lines, bearings and between surfaces when the machines are not in operation. It must be mentioned that the presence of mineral oil, which will form a skin over fluids in settling tanks, etc., can also give rise to septic conditions during shut-downs; it is possible to minimise this effect by the injection of a slow stream of air into the fluid. Although chemical germicides are normally used for bacteria control they are no substitute for aeration and the prompt removal of swarf.

THE TREATMENT OF REJECT CUTTING FLUIDS

When considering the treatment of oil-based cutting fluids, the most usual method is to allow time for the fluid to settle out and then to place the oil in drums for eventual disposal to a waste-oil merchant.

The treatment of rejected emulsified oils is a more complex matter. The treatment that is acceptable to most authorities is to break the emulsion and to separate the oil phase; it is possible to do this by allowing the fluid to stand for a lengthy period but, because of the time element, chemical methods are usually used. These methods are based on de-activating the emulsifier compound which forms the link or bridge between the oil and the water.

An emulsified cutting fluid is formed by dispersing oil in the form of minute droplets in water, making a temporary mixture. Such a mixture is thermo-dynamically unstable and will revert to fully separated phases by coalescence and sedimentation but stabilising agents will interpose a film in the interface between the emulsified-oil droplets and the continuous phase, i.e. the water. Surface tension of the oil promotes cohesion but this is opposed by the surface potential of the emulsifier-coated droplets which causes them to tend to repel others which are similarly charged. Emulsifying agents used with cutting fluids are usually anion-active agents such as sodium, potassium or amino soaps, sulphonated oils and sulphonated alcohols, the most suitable type depending on a number of factors including the hardness of the water.

The breaking of emulsions is based upon reducing the electrical charge on the oil particles thus counteracting the effect of the emulsifier. Physical methods of breaking can be used and these include electrical methods, the addition of oil, centrifuging, ultrasonic agitation, etc. When the breaking is done by chemical methods, electrolytes are added to the emulsion and this reduces the number of free charges on particle surfaces. Alternatively, acids can be added which will convert soaps to fatty acids. Whichever method is used, when the emulsifier has been broken, the oil droplets coalesce into large drops which may be removed by gravity separation or skimming. If the breaking stage is followed by a clarifying process then chemical coagulation and flocculation will aid the complete removal of the oil. Choice of the most suitable and economic process depends on a number of factors including degree of dilution, type of emulsifier and volume of fluid. Emulsions of high concentration can be broken easily but it is more difficult to break dilute emulsions.

As a general guide it may be taken that the following applies:

1 An acid inorganic salt is usually more effective than an acid and a salt.

2 A salt of a divalent or trivalent metal (iron, aluminium, calcium) is more effective than a salt of a monovalent metal (sodium, potassium).

3 In the presence of chemical flocculants, emulsion breaking is

most effective when the acidity is such that a floc is just formed.

4 In the case of small volumes the results of a simple batch breaking step followed by decantation may be satisfactory, but when large volumes are involved continuous processes using multi-stage operation may be necessary.

The actual breaking of the emulsion will usually be carried out after the free oil and heavy sediments have settled out. After adding the electrolyte the mixture is allowed to stand for a short period and then the oil may be removed; if it can be arranged, a further short period of aeration after mixing may assist emulsion breaking. The resultant aqueous phase will be reasonably free from oil and is usually suitable for direct discharge into a sewer; however, if acids, alkalis or acid salts are used to break the emulsion these may need to be neutralised before discharging.

Figures 11.1 and 11.2 shows two plants suitable for dealing with reject emulsions and oily waters.

THE ROLE OF THE MACHINE-TOOL DESIGNER

Although the problem of the disposal of spent cutting fluid is primarily one for the user of machine tools, the machine-tool designer is also intimately concerned with this problem and particularly so with the following aspects:

1 To design the machine so that any cutting-fluid containers within the structure are readily available for cleaning purposes and are of such a shape that cleaning is made easy.

2 To arrange the design so that equipment for the preliminary cleaning of the cutting fluid is built into the machine, thus prolonging the life of the fluid and retarding bacterial growth.

3 To design the machine so that the dangers of pollution are minimised.

SUMMARY

The constantly growing knowledge of pollution and improved methods of waste treatment are leading to progressively more

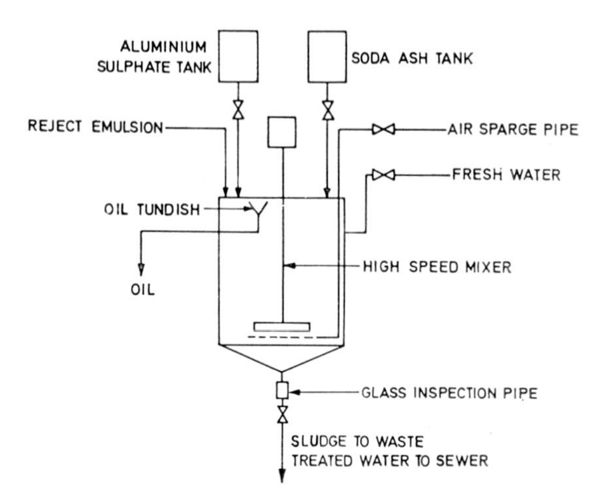

Fig 11.1 Batch tank for treatment of reject emulsions

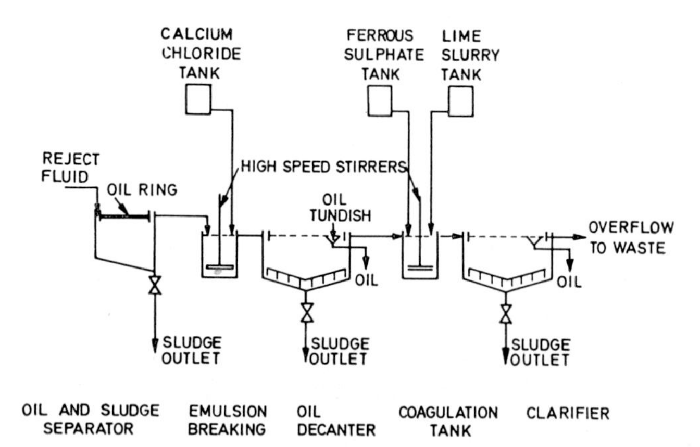

Fig 11.2 Simplified flow diagram of plant for continuous treatment of oily waters

rigid enforcement of recent anti-pollution legislation. Unauthorised pollution is a criminal offence. It is quite likely that future effluent limits will approach drinking-water standards and suitable waste treatment must be accepted as a production cost. Thus industry must aim to keep within the bounds of the law in as economic a way as possible.

Oil in effluents from machining operations presents an increasingly serious pollution problem but this can be reduced by the following:

1 adequate recycle fluid-treatment plants to reduce the volume rejected;

2 designing machines to facilitate good housekeeping and to reduce accidental spillages;

3 installing treatment plants for waste fluids.

The recycle and waste-fluid treatment plants cannot be considered in isolation but must be regarded as integral parts of the production plant.

The initial responsibility for minimising pollution lies with the machine designer, secondly with the purchaser and thirdly, but by no means least, with the machine operator. Greater co-ordination between designers and users is essential to ensure that this situation is thoroughly understood and appreciated.

Some waste fluid is inevitable and it must be treated before discharge. Treatment methods based on settling, emulsion breaking, coagulation and pressure flotation can result in acceptable contamination levels. These processes are made easier by segregating cutting fluids and oily wastes. Optimum treatment conditions vary according to the type of effluent but little processing information is available on individual cutting fluids.

It is quite likely that future cutting fluids will contain an increasing number of additives and special waste-treatment processes will have to be developed to cope with this situation.

APPENDIX

A GLOSSARY OF TERMS USED IN CONNECTION WITH THE HANDLING, PROCESSING AND TRANSPORTING OF SWARF

SECTION 1 GENERAL

Machining
The operation of changing the shape and size of a workpiece by removing a portion of its bulk or volume by the use of some form of cutter employed in a suitable apparatus, usually a machine tool.

Swarf
The name given to the material removed in changing the shape and size of a piece of material during a machining operation. There are three main types:
1 *Bushy swarf* is normally a tangled formation of long helices of the removed material.
2 *Broken chips* are pieces of material from broken helices, often one full turn in length, or chips or splinters.
3 *Dust and sludge* are particles of material usually less than $\frac{1}{16}$ in diameter. These are classified as dust from dry machining and as sludge when coolant or cutting oil is used.

Machine periphery
The boundary of a machine tool within which, in the interests of health and good housekeeping, all swarf and coolant should be contained during machining operations.

Unit system
This is so called when the swarf-removal arrangements are designed to deal only with the swarf produced by a single machine.

Central system
A system which caters for the collection of swarf from a complete department or a number of departments, and removes it to a central location for processing and disposal.

Tractor or tug unit
A unit which does not itself carry a load but is used for towing one or more non-mechanical trailers or mobile skips. It may be powered by battery, diesel or petrol engine.

Culvert, trough or channel
This is usually a trench cut into the workshop floor and along which swarf is conveyed mechanically or hydraulically to a central gathering point.

Pit or sump
A collecting point for the swarf transported from the machine tools by any means, including manual.

Micron
A sub-unit of length in the metric system—one millionth of a metre or 0·000039 in.

SECTION 2 COOLANTS AND CUTTING OILS

Coolant
Generally, this is water or a mixture of water and some other ingredient(s) used to cool and lubricate the cutting tool and workpiece during machining operations. The additions may include emulsified mineral oil or chemicals. Coolants are also referred to as cutting compounds or cutting fluids. Paraffin is another type of coolant frequently used for honing operations.

Cutting oil
This is usually based on a mineral oil but will probably also contain vegetable and mineral oils with other additives.

Coolant cleanliness
The degree of purity that should be maintained in the coolant so that cutting is not adversely affected and operator health is not jeopardised.

Dirty coolant
After or during a machining operation, this is the condition of the coolant as it is delivered from the machine tool to the filter, separator or other device for cleaning.

Clean coolant
The condition of the coolant following treatment which has removed the solid particles and rendered the fluid suitable for further use.

Polluted coolant
The description of a coolant that has become contaminated by bacteria or other matter, rendering it unfit for further use.

Tramp oil
Any oil, such as hydraulic or lubricating, which becomes mixed with the coolant and which is incompatible.

Complementary oils
A new range of oils that are suitable for both hydraulic and lubricating purposes and, in limited quantities, are compatible with many cutting fluids.

Effluent
The fluid portion of a mixture which remains after treatment by filtration, separation or some other means, and is often of no further use.

Spillage
That fluid which may fall to the floor outside the periphery of the machine tool during cutting operations or during swarf removal.

Seepage
That fluid which may ooze out, trickle or leak from the swarf during handling, processing, storing, loading or transporting.

Emulsion
A dispersion of fine liquid particles in a liquid stream, in which they do not dissolve but are held in suspension.

SECTION 3 FILTERS AND SEPARATORS

(a) FILTER

An apparatus used for separating suspended, undissolved particles of solids from liquids by means of a porous medium.

Filter media
The porous materials used for separating solids from liquids.

Filter cake
The name given to the accumulation of solids deposited on the filter medium during the filtration process.

Filter pre-coat
The depositing of a chemically inert material on a filter medium.

Coagulation
The action of very fine particles of colloidal size adhering directly to each other. The electrical charges on the particle surfaces need to be neutralised by a suitable additive.

Flocculation
The formation of open clusters of particles brought about by molecules of reagent acting as bridges between separate suspended particles.

Gravity filter
The contaminated coolant is fed into the chamber where, under gravity, the liquid passes through a porous medium which is suitably supported.

Pressure filter
Coolant, under pressure from a pump, is forced through a porous medium contained in a suitable vessel.

Vacuum filter
The liquid portion of a contaminated mixture is forced through a porous medium into a vacuum chamber. The depression in the chamber is produced either by the coolant pump or a separate vacuum pump.

(b) SEPARATORS

Devices where some external forces are applied in order to achieve separation.

Settling chamber
A tank or sump in which a liquid/solid mixture rests or flows sufficiently slowly to allow solid particles to settle out of the mixture and be deposited on the bottom.

Flotation separator
A device designed to introduce large quantities of air bubbles into a liquid/solid mixture; the bubbles attach themselves to the particles of contaminant and float them to the surface.

Magnetic separator
Equipment in which magnetic forces are used to separate ferrous contaminant from a liquid/solid mixture.

Centrifuge
An apparatus wherein high-speed rotation creates considerable centrifugal force which brings about the separation.

Basket centrifuge
A centrifuge having a solid-wall bowl which is open at one end.

Tubular-bowl centrifuge
A centrifuge with liquid/solids input through a nozzle at the bottom of the tubular bowl. Liquid phases are discharged from the top.

Disc-bowl centrifuge
A centrifuge which has a solid-wall bowl containing a disc stack within the bowl. Solids are retained within the bowl whilst liquid phases are discharged from the top of the machine.

Self-cleaning disc-bowl centrifuge
This is similar to the disc-bowl model except that the solids are discharged automatically from the periphery of the bowl.

Multi-chamber centrifuge
A centrifuge having a solid-wall bowl containing a number of chambers. Only liquid/solid separation is possible with this type which has to be manually cleaned.

Decanter
A type of centrifuge, usually operating horizontally, often equipped with

an internal scroll arrangement and generally used for continuous throughput.

Cyclone
This is a separation unit based on the varying densities of materials carried in an air stream. The tangential entry of the airstream creates a rotational motion which throws the solids against the sides and allows them to fall for discharge from the bottom of the cone. The cleaned air forms a central vortex and discharges from the top of the cyclone.

Hydrocyclone
This piece of equipment embodies all the characteristics of the cyclone but is capable of dealing with solid/liquid mixtures, centrifugal force again causing the separation.

SECTION 4 CONVEYORS

Conveyor
A horizontal, inclined or vertical device for moving or transporting materials in granular or packaged forms.

Scraper, drag bar or drag-link conveyor
A conveyor having one or more chains equipped with scraper bars and operating in a trough.

Belt conveyor
A conveyor using a moving belt as the conveying medium. The belt is usually driven by a drum at one end, passing over a free-running drum at the other end. The upper portion of the belt may be supported by free-running idlers or suitable flat surfaces. This type of conveyor can be arranged for horizontal or inclined travel, the angle of slope depending on the character of the goods conveyed and the type of belt surface.

Bucket elevator
An elevator, generally totally enclosed, for powder or granular materials, consisting of suitably shaped buckets mounted at predetermined pitches on an endless belt. In place of the belt, one or more chains may be used for carrying the buckets. This type of elevator is usually employed for lifting materials vertically or at a steep angle.

Reciprocating conveyor

Basically, this is a hydraulically operated drag conveyor comprising a shaft fitted with fins to move the swarf along, and located in a U-shaped trough. Normally, the linear motion is imparted by a hydraulic ram having a stroke of approximately five feet. The movement is slow, the bushy swarf and cuttings being carried forward past tines projecting from the sides of the trough which hold the swarf against the return stroke of the ram.

Vibratory conveyor

A conveyor which consists of a metallic or plastic trough or tube, flexibly mounted, to which is imparted a vibratory movement, usually of sawtooth-wave form, the amplitude and frequency being made variable. The material travels along the trough due to this movement.

Slat conveyor

An elevator/conveyor consisting of a belt or one or more endless chains, suitably supported, to which spaced slats are attached for carrying the loads.

Screw conveyor

A conveyor in which the conveying element is in the form of a solid helix with either constant or varying pitch, attached to a central rotating shaft, the whole being enclosed in a U trough or tube, the material being moved along by the action of the helix. In some similar conveyors, the solid helix may be replaced by paddle blades of various shapes attached to the central shaft in a helical pattern.

Magnetic conveyor

TYPE A Magnetic assemblies move under a stainless-steel skin. The mixture enters the hopper and passes under a baffle-plate where separation take place; the cleaned coolant is returned over the baffle and the magnets carry away the particles of ferrous swarf.

TYPE B A moving belt passes over either stationary or moving magnetic assemblies; operation generally is as Type A.

Both types are highly efficient for the separation of light ferrous swarf particles, the efficiency depending upon the flow rate and viscosity of the contaminated coolant.

Hydraulic conveying

In this arrangement, swarf is transported by the volume and velocity of fluid, both being sufficiently great to support and carry the particles along. This system will carry chips or sludge but has limitations when the swarf is of bushy form.

Velocity trench
This type of trench is used only for the conveyance of swarf by hydraulic means. As transportation depends upon the swarf particles remaining in suspension, jets of liquid may be introduced at strategic points in order to maintain the velocity of the liquid carrier.

Pneumatic conveying
In this type, particles of swarf are drawn into a duct and transported by air moving with high velocity. This system will remove small chips and dust from the immediate vicinity of the cutting tool providing the moisture content is very low.

SECTION 6 SWARF-PROCESSING EQUIPMENT

Crusher or hammer mill
A machine into which bushy swarf is fed to be broken into smaller pieces making for easier handling, storing and transporting.

Briquetting machine
A machine into which swarf is fed and is compressed into a compact mass, usually of small dimensions.

Packaging
The forming of coarse swarf or chippings into bales or briquettes by compressing such swarf in a machine built for that purpose.

INDEX